DATE DUE

MY 0 4 '74		
OC 17 '77		
AP 20 '78		
AG 03 '78		
MR 17 '20		

Demco, Inc. 38-294

COLOMA PUBLIC LIBRARY
COLOMA, MICHIGAN 49038

WORLD BOOK'S LIBRARY OF NATURAL DISASTERS

FORCES OF NATURE

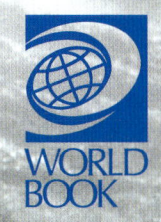

WORLD BOOK

a Scott Fetzer company
Chicago
www.worldbookonline.com

World Book, Inc.
233 N. Michigan Avenue
Chicago, IL 60601
U.S.A.

For information about other World Book publications, visit our Web site at **http://www.worldbookonline.com** or call **1-800-WORLDBK (967-5325)**.

For information about sales to schools and libraries, call **1-800-975-3250 (United States)**; **1-800-837-5365 (Canada)**.

2nd edition
© 2009 World Book, Inc. All rights reserved. This book may not be reproduced in whole or in part in any form without prior written permission from the publisher.

WORLD BOOK and the GLOBE DEVICE are registered trademarks or trademarks of World Book, Inc.

The Library of Congress has cataloged an earlier edition of this title as follows:

Forces of nature.
    p. cm. -- (World Book's library of natural disasters)
    Summary: "A discussion of major types of natural disasters, including descriptions of some of the most destructive; explanations of these phenomena, what causes them, and where they occur; and information about how to prepare for and survive these forces of nature. Features include an activity, glossary, list of resources, and index"--Provided by publisher.
    Includes bibliographical references and index.
    ISBN 978-0-7166-9806
    1. Weather--Juvenile literature. 2. Climatology--Juvenile literature. 3. Geology--Juvenile literature. 4. Communicable diseases--Juvenile literature.
I. World Book, Inc.
QC981.3.F67 2008
363.34--dc22
                        2007006186

This edition:
ISBN: 978-0-7166-9822-7 (Forces of Nature)
ISBN: 978-0-7166-9817-3 (set)

Printed in China
1 2 3 4 5 12 11 10 09 08

Editor in Chief: Paul A. Kobasa

Supplementary Publications
    Associate Director: Scott Thomas
    Managing Editor: Barbara A. Mayes
    Editors: Jeff De La Rosa, Nicholas Kilzer, Christine Sullivan, Kristina A. Vaicikonis
    Researchers: Cheryl Graham, Jacqueline Jasek
    Manager, Contracts & Compliance (Rights & Permissions): Loranne K. Shields

Graphics and Design
    Associate Director: Sandra M. Dyrlund
    Associate Manager, Design: Brenda B. Tropinski
    Associate Manager, Photography: Tom Evans
    Designer: Matt Carrington

Production
    Director, Manufacturing and Pre-Press: Carma Fazio
    Manager, Manufacturing: Steven Hueppchen
    Manager, Production/Technology: Anne Fritzinger
    Proofreader: Emilie Schrage

Product development: Arcturus Publishing Limited
    Writer: Philip Steele
    Editors: Nicola Barber, Alex Woolf
    Designer: Jane Hawkins
    Illustrator: Stefan Chabluk

Acknowledgments:

AP Photo: 24 (Rich Pedroncelli).

Corbis: 4 (A.J. Sisco), 7 (William Campbell), 10 (Corbis), 13 (Jim Zuckerman), 15 (Nigel Francis), 25, 31 (Roger Ressmeyer), 26 (Peter M. Wilson), 33 (Goran Tomasevic/ Reuters), 35 (Kimimasa Mayama/ Reuters), 36 (Ulises Rodriguez/ epa), 38 (Bettmann), 43 (Pallava Bagla).

NASA: 12 (Jeff Schmaltz, MODIS Rapid Response Team, NASA/ GSFC).

Magnum Photos: cover/ title page (Jean Gaumy).

Science Photo Library: 5 (Aaron Johnson & Brooke Tabor/ Jim Reed Photography), 8, 30, 32, 34 (Gary Hincks), 9 (George Post), 17 (NASA), 18 (Simon Fraser), 19, 42 (Sinclair Stammers), 20 (Rev. Ronald Royer), 23 (Paolo Koch), 27 (Michael Gilbert), 29 (David Parker), 37 (W. Bacon), 39 (NIBSC), 40 (Eye of Science), 41 (National Library of Medicine).

TABLE OF CONTENTS

- **Dangerous forces** — 4-5
- **Forces of weather** — 6-27
  - What is weather? — 6-7
  - Water in the air — 8-9
  - Atmospheric pressure and wind — 10-11
  - Storms — 12-13
  - Earth's climates — 14-15
  - Why climates vary — 16-17
  - Climate and the environment — 18-19
  - Ocean currents — 20-21
  - Seasons — 22-23
  - Unseasonable weather — 24-25
  - Climate change — 26-27
- **Forces of geology** — 28-37
  - Earth's layers — 28-29
  - Volcanoes — 30-31
  - Earthquakes — 32-33
  - Tsunamis — 34-35
  - Landslides, mudslides, and avalanches — 36-37
- **Forces of disease** — 38-43
  - Microorganisms — 38-39
  - Bacteria and viruses — 40-41
  - Parasites and fungi — 42-43
- **Activity** — 44
- **Glossary and Additional resources** — 45-47
- **Index** — 48

Glossary There is a glossary of terms on pages 45-47. Terms defined in the glossary are in type **that looks like this** on their first appearance on any spread (two facing pages).

Additional resources Books for further reading and recommended Web sites are listed on page 47. Because of the nature of the Internet, some Web site addresses may have changed since publication. The publisher has no responsibility for any such changes or for the content of cited sources.

DANGEROUS FORCES

The **planet** Earth is our home. Not only is it our home, all the living things we know of are on Earth. We think of this home as a safe place. But there are forces operating on Earth that can make our planet very dangerous at times.

The **weather** that makes Earth liveable can also bring harm to Earth's inhabitants. The rains that allow crops to grow can also bring **floods** and **ice storms** that cause devastation. The winds that cool us can also form such storms as **tornadoes** and spread **wildfires.**

Hurricane Katrina flooded large areas of New Orleans in August 2005.

DANGEROUS FORCES

The forces that formed and continue to form the Earth we stand on can be destructive forces. **Volcanoes** that produce new **crust** can also burn entire cities that are in the path of an **eruption**.

Some kinds of **bacteria** *(bak TIHR ee uh)* are useful. For example, bacteria that live in our bodies help us to digest our food. Other kinds can cause serious diseases that can wipe out thousands of people in a single outbreak. **Cholera** *(KOL uhr uh)*, for example, is a highly dangerous, water-borne, bacterial disease.

The forces of Earth can be deadly, destructive forces. They can lead to suffering and large loss of life. They can lead to disasters.

Forces of Nature explains the natural forces that often result in disaster.

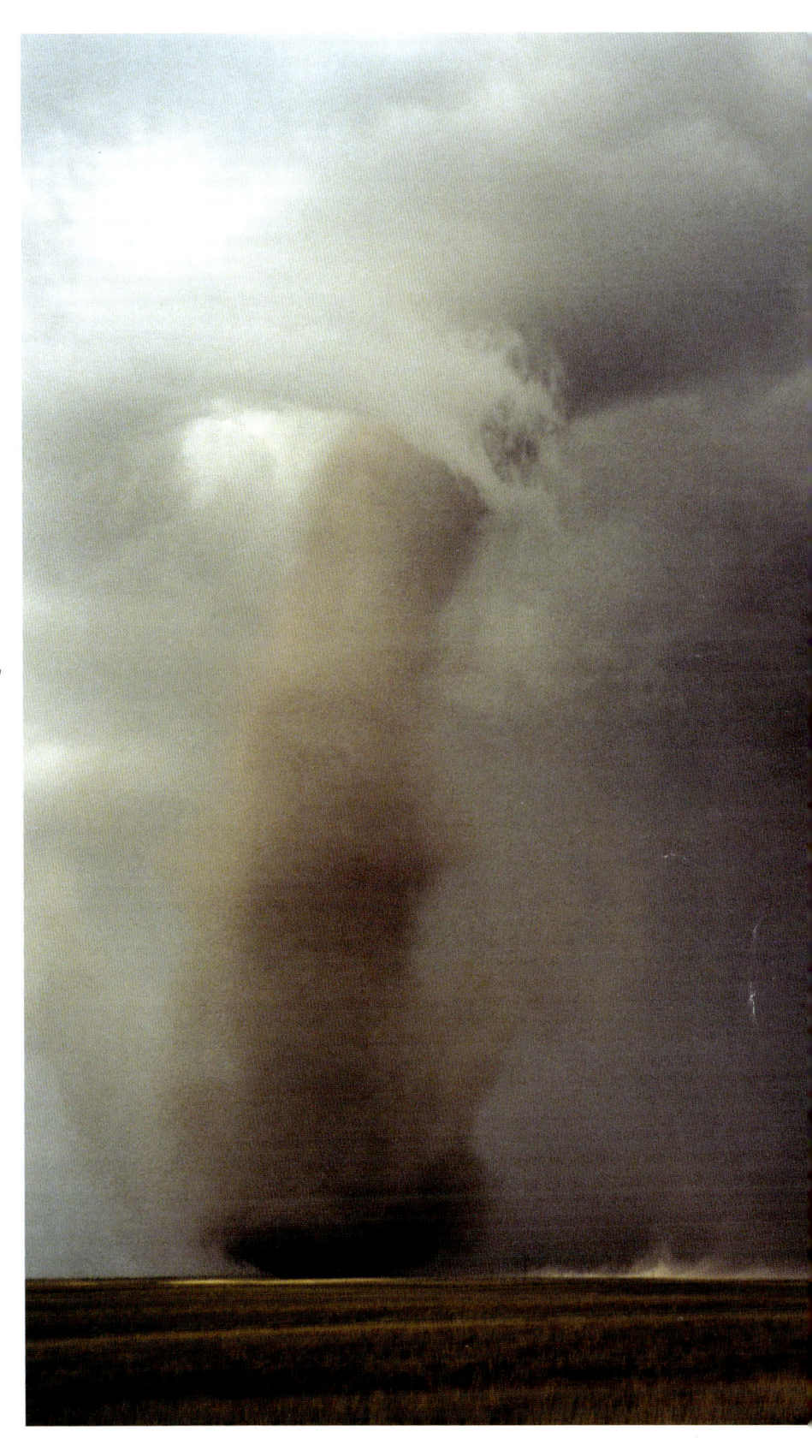

Deadly storms, such as tornadoes, can be very destructive.

FORCES OF WEATHER

WHAT IS WEATHER?

Weather is the state of the **atmosphere** *(AT muh sfihr)*—the blanket of air that surrounds Earth—at a particular place and time. If the air is moving quickly, it is windy. If there are clouds, it could be rainy or snowy. The temperature of the air can be hot or cold. The weather in a particular place may change from day to day and sometimes even from hour to hour. If the weather is severe, it can sometimes cause serious problems. We measure and describe the weather in many different ways:

- Wind is the movement of air across the surface of Earth. Wind can be moving so fast as to cause serious damage or so slowly and gently that we can hardly feel it.
- **Precipitation** *(prih SIHP uh TAY shuhn)*—such as rain, snow, and hail—forms in clouds in the sky.

The troposphere is the lowest layer of the atmosphere. Nearly all of Earth's weather—including clouds, rain, and snow—occurs in the troposphere.

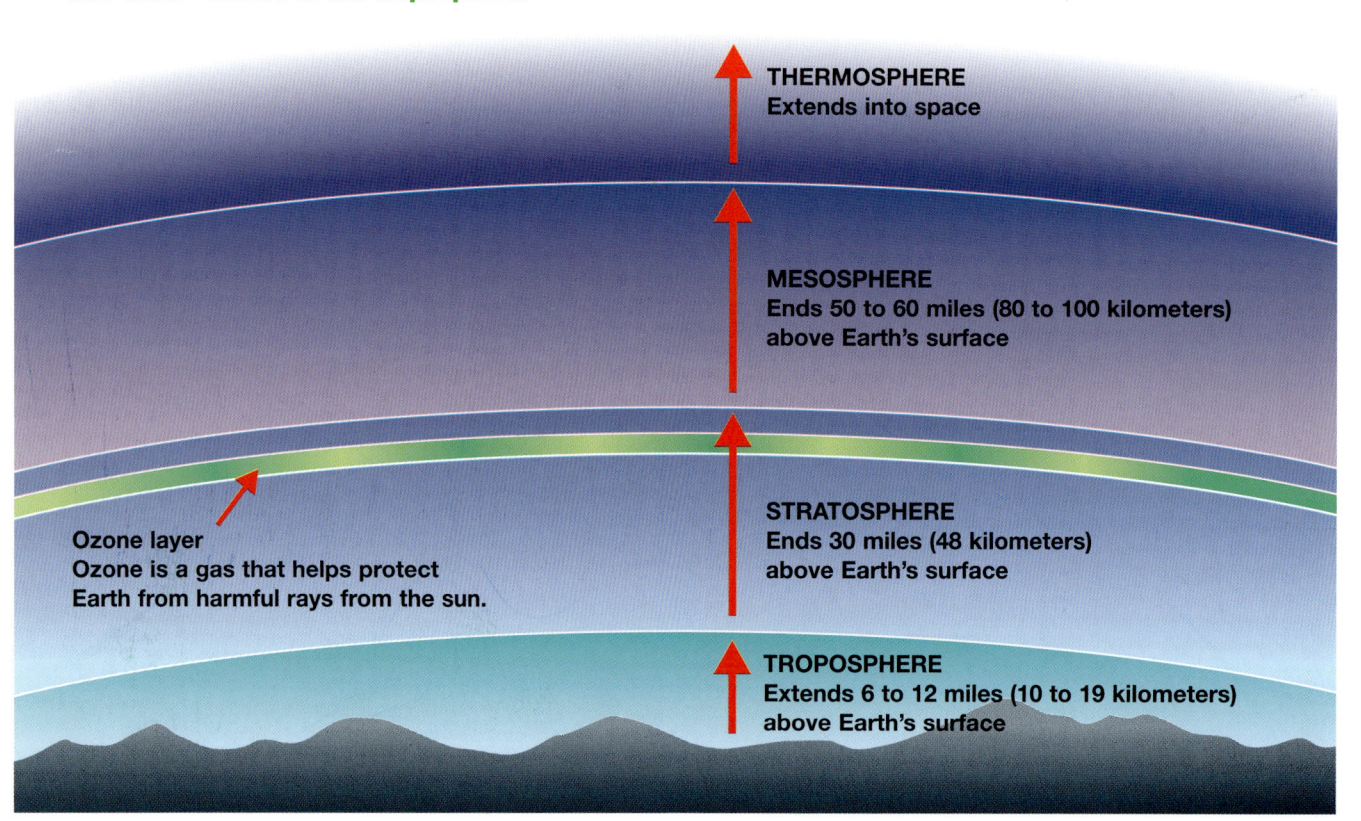

THERMOSPHERE
Extends into space

MESOSPHERE
Ends 50 to 60 miles (80 to 100 kilometers) above Earth's surface

STRATOSPHERE
Ends 30 miles (48 kilometers) above Earth's surface

Ozone layer
Ozone is a gas that helps protect Earth from harmful rays from the sun.

TROPOSPHERE
Extends 6 to 12 miles (10 to 19 kilometers) above Earth's surface

WHAT IS WEATHER? **7**

- Temperature is a measure of how hot or cold the air is.
- Visibility is the condition of the atmosphere with reference to the distance at which things can be clearly seen. Visibility is often poor when there is fog, mist, or rain.

Living with the weather

People learn to cope with the kinds of weather that are expected in a particular place. For example, in Canada, the northern United States, Russia, and Scandinavia people have learned how to handle very cold temperatures and heavy snow. Their houses are built to keep them warm, and they use snowplows and other equipment to keep roads clear of snow.

A snow plow clears deep snow in Yellowstone National Park, in preparation for the opening of the park for the summer season.

Extreme weather

When the weather does unexpected things, it can cause many kinds of problems. If there is a lot of rain, rivers may overflow, causing **floods.** Floods can destroy buildings and sometimes cause great loss of life. If there is not as much rain as usual, there may be a **drought.** During times of drought there is often not enough water for crops to grow or for animals to drink, and plants and animals may die.

THE ATMOSPHERE

Our weather takes place in the atmosphere. The layer of atmosphere closest to the surface of Earth is called the **troposphere** (TROH puh sfihr). This is where the gases that make up the atmosphere are thickest. The troposphere contains 99 percent of the water in the atmosphere, and it is where clouds form. Farther away from Earth's surface, the gases become thinner, in layers called the **stratosphere** (STRAT uh sfihr), the **mesosphere** (MEHS uh sfihr), and the **thermosphere** (THUR muh sfihr).

FORCES OF NATURE

WATER IN THE AIR

Precipitation happens as a part of the **water cycle,** or hydrologic cycle. This cycle is the continuous movement of water rising from the oceans to the air, falling from the air to the land, and making its way back to the oceans again.

The water cycle

When the sun shines on the oceans or other areas of water, its heat causes the water to **evaporate** and become invisible **water vapor.** This vapor rises, and as it reaches cooler air, it **condenses** into clouds of tiny water droplets, which are blown across the sky by the wind. The droplets cool even more in the upper air and merge to become larger drops, which fall to Earth as rain.

In the water cycle, water moves between the surface of Earth and the atmosphere.

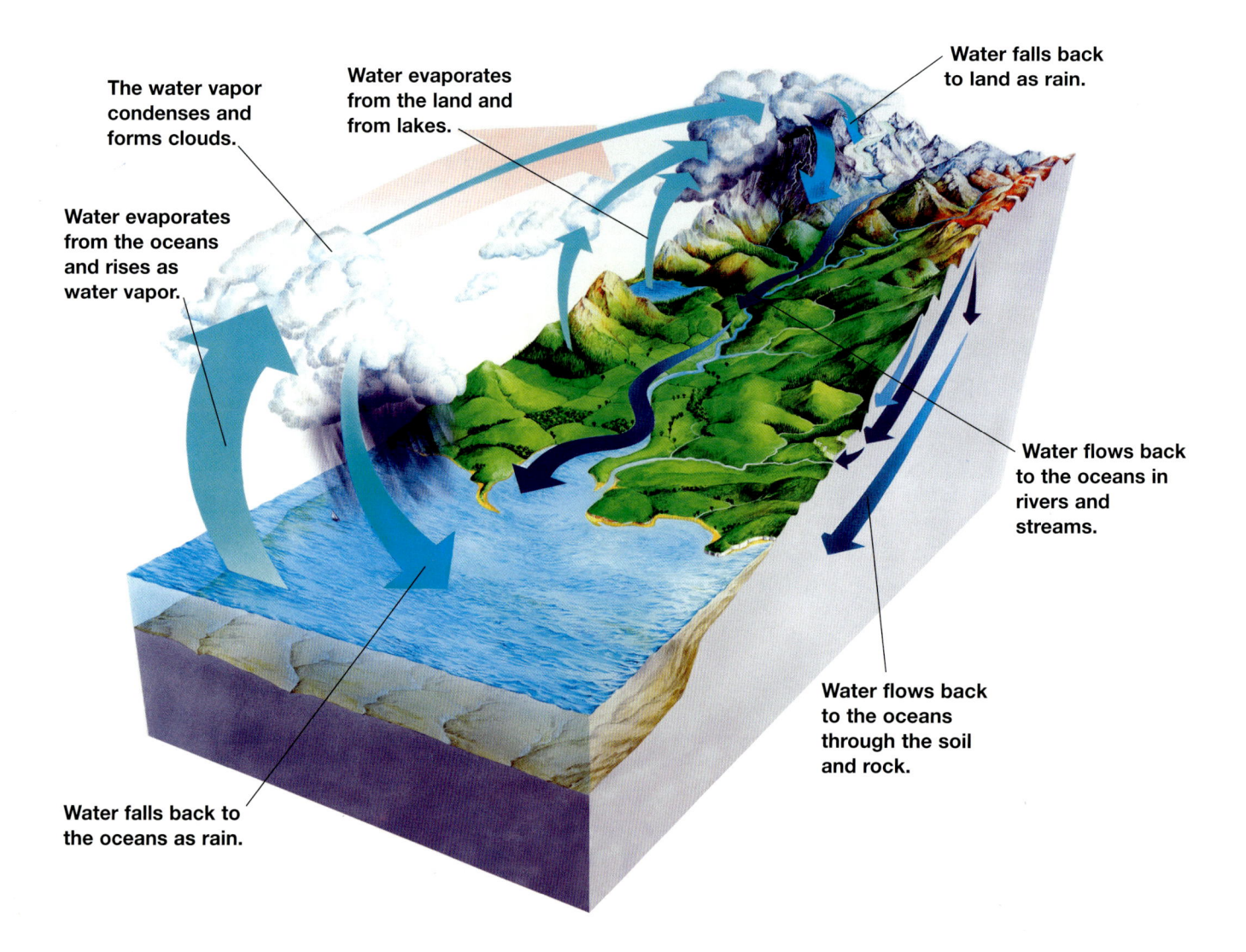

- The water vapor condenses and forms clouds.
- Water evaporates from the land and from lakes.
- Water falls back to land as rain.
- Water evaporates from the oceans and rises as water vapor.
- Water flows back to the oceans in rivers and streams.
- Water flows back to the oceans through the soil and rock.
- Water falls back to the oceans as rain.

WATER IN THE AIR 9

Sometimes, especially during a thunderstorm, raindrops freeze before they reach the ground, forming lumps of ice called hail. Precipitation may also fall as snow. With snow, the tiny droplets in the clouds freeze to form ice crystals, which then fall to the ground.

Cloud cover

Earth's temperature can be influenced by clouds. Sometimes, cloudy days are cooler than clear days because the clouds reflect much sunlight back into space. At night, clouds often have an opposite influence on the temperature of the air near Earth's surface. Earth cools by giving off heat toward space. Clouds intercept much of this heat and send it back toward the ground. For this reason, most cloudy nights are warmer than clear nights.

Clouds are usually high above the surface of Earth, but water vapor in the air can sometimes condense close to Earth's surface. We call this ground-level cloud fog or, if it is not very thick, mist. When the water vapor condenses onto the ground and on objects as drops of water, it forms dew. If water vapor near the ground freezes into ice crystals, it is transformed into frost.

Thick, white frost has formed on the ground and on these trees.

FOG OVER THE DESERT

One of the strangest **climates** in the world is found along the coast of Namibia, in southwest Africa. A few miles inland, daytime temperatures in the hot, dry Namib **Desert** are very high. But the air is much cooler along the coast because of the cold Benguela ocean **current** that flows north through the Atlantic Ocean from Antarctica. The cool temperatures make moisture in the air condense, so that thick fog often forms over the desert close to the ocean, though rain hardly ever falls there.

ATMOSPHERIC PRESSURE AND WIND

The air in the **atmosphere** presses down onto Earth's surface. This is **atmospheric pressure.** The pressure in the upper layers of the atmosphere is not as great as in the layers nearer to Earth's surface because there is less air pressing down in these higher regions of the atmosphere. Atmospheric pressure can also vary at Earth's surface. Warmer air is less dense than cold air, so warm air exerts less pressure. Air along the surface of Earth tends to move from high-pressure to low-pressure areas, blowing as wind.

General circulation

The sun heats Earth's surface unevenly (see page 22). The sun shines directly on areas at the **equator.** When it shines on areas further to the north or south, it shines at an angle because of the tilt of Earth's **axis.** So, the sun warms places near the equator more than places farther north or south. This warm, low-pressure air at the equator rises. As warm air rises, air from cooler areas moves in to take its place. Scientists call these movements "general circulation."

A Florida beach is pounded by high winds during Hurricane Georges in August 1998.

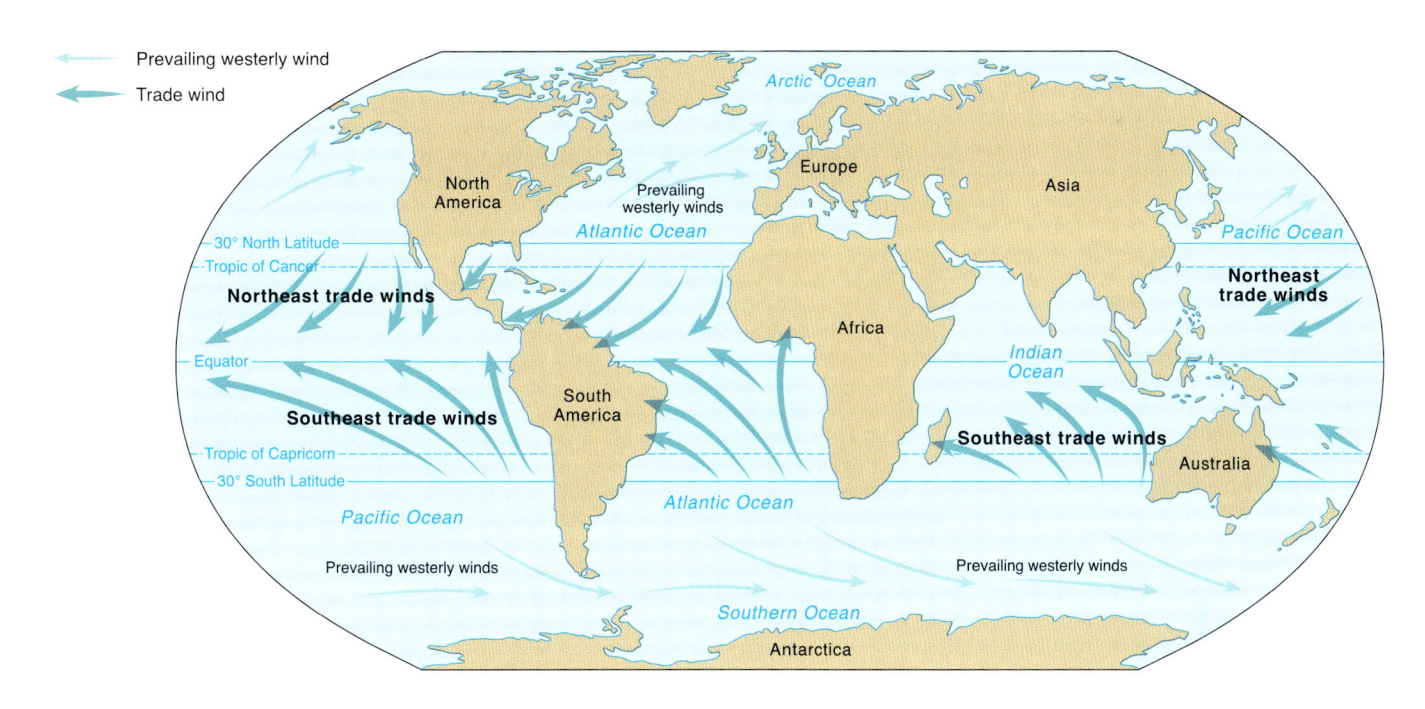

ATMOSPHERIC PRESSURE AND WIND 11

Prevailing winds

General circulation produces the kinds of winds that are characteristic of a given area. These are known as **prevailing** *(prih VAY lihng)* **winds.** Prevailing winds in the tropics are known as trade winds.

The spinning motion of Earth on its axis has an effect on the prevailing winds. North of the equator, winds are forced to the right. South of the equator, winds are forced to the left. This is called the **Coriolis** *(KAWR ee OH lihs)* **effect,** named after French scientist Gaspard Gustave de Coriolis (1792-1843), who first described the effect in 1835.

Jet streams

The Coriolis effect is just one of the factors affecting the direction in which the wind blows. Another is the existence of fast-moving air **currents** high in Earth's atmosphere, called **jet streams.** These currents help form and strengthen areas of low pressure and so have an important effect on our weather.

Trade winds are strong prevailing winds that occur mainly in the tropics. The direction of the trade winds above and below the equator demonstrates the Coriolis effect.

LOCAL WINDS

Some winds happen frequently though they are not prevailing winds. One example is known as a katabatic *(KAT uh BAT ihk)* wind. This kind of wind occurs when cold air gathers in mountain valleys and then blows down from the mountains to the lands below. One of the most famous katabatic winds is the mistral *(MIHS truhl)*, which often blows for 100 days a year in southeast Europe.

STORMS

The wind is a powerful force. High winds at sea produce huge waves that can wreck ships and damage coastal settlements. In **deserts,** the wind can whip up sandstorms and create huge banks of sand—called dunes—that migrate over time.

A satellite image over the Gulf of Mexico shows Hurricane Katrina approaching the Gulf Coast of the United States in August 2005.

Blizzards

A **blizzard** is a fierce snowstorm with high winds and low temperatures. Blizzards occur when cold **Arctic** air advances toward an area of warm, moist air. The place where they meet is called a cold **front.** The warm air is forced upward, cooling as it rises. The cooling air can no longer hold its moisture, which drops to the ground as snow.

Hurricanes and other tropical cyclones

Some of the most powerful winds occur during storms that form over **tropical** oceans. Winds created by these tropical cyclones can reach speeds of up to 200 miles (320 kilometers) per hour. The most severe tropical cyclones are called **hurricanes** when they happen over the North Atlantic Ocean, the Caribbean Sea, the Gulf of Mexico, or the Northeast Pacific Ocean. They are called **typhoons** if they occur in the Northwest Pacific Ocean. Near Australia and in the Indian Ocean, they are simply referred to as severe **tropical cyclones.** All three terms—hurricane, typhoon, and tropical cyclone—refer to the same type of storm. The storms just go by different names in different regions.

STORMS   13

Tropical cyclones start over warm oceans as the warm water **evaporates** into the air above. The warm, moist air rises to create an area of low pressure beneath. Air rushes into the low-pressure area, and winds start to swirl around a central area of much stiller air, called the eye of the storm. If these winds reach 74 miles (119 kilometers) per hour, the storm is considered to be a tropical storm. Tropical storms move across the surface of an ocean and sometimes across land, often causing enormous damage. For example, Hurricane Katrina hit the United States in late August 2005. As a result of the storm, at least 1,800 people were killed and large sections of New Orleans, Louisiana, were destroyed.

Tornadoes

Like hurricanes, **tornadoes** *(tawr NAY dohz)* are areas of spinning wind. However, most tornadoes are smaller than 1,600 feet (480 meters) across, while a hurricane may be hundreds of miles across. The winds of a tornado are faster than those of a hurricane, sometimes exceeding 300 miles (480 kilometers) per hour.

RAINING FROGS AND FISH
There are occasionally strange stories of frogs or fish raining down from the sky. These rains are the result of tornadoes that have formed over water, creating **waterspouts.** Fish or frogs may be drawn out of the water and up into the clouds and then dropped many miles away.

The wind speeds of a tornado, such as this one in the rural Midwest of the United States, generally exceed those of a hurricane.

FORCES OF NATURE

EARTH'S CLIMATES

Climate is the kind of **weather** usually experienced in a particular place, averaged out over a long time. For example, in a dry climate rain rarely falls. A cold climate has few, if any, days with warm temperatures.

The different kinds of climates on Earth can be divided into 12 main groups:

- **Tropical** wet climate
  Temperatures rarely drop below 64 °F (18 °C) and there is heavy rainfall for much of the year.
- Tropical wet and dry climate
  Temperatures are warm for most of the year, but rain falls only during certain months.
- **Desert** climate
  There is very little rain; daytime temperatures in a desert may be hot, but nights are cold since there are no clouds to keep the air warm.

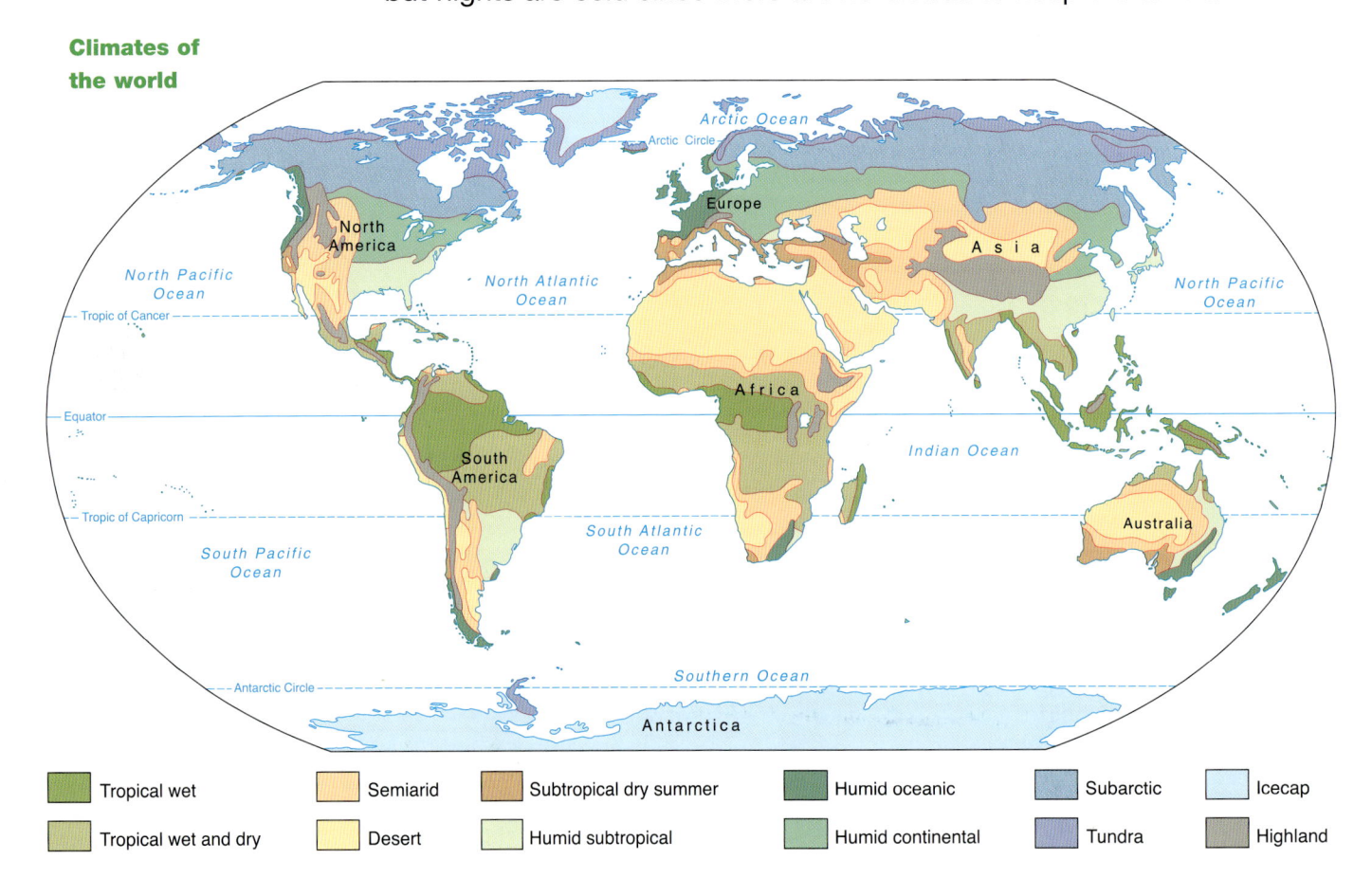

Climates of the world

Tropical wet	Semiarid	Subtropical dry summer	Humid oceanic	Subarctic	Icecap
Tropical wet and dry	Desert	Humid subtropical	Humid continental	Tundra	Highland

- **Semiarid** climate

  The semiarid climate borders desert climates and has similar temperatures, but it is not as dry as a desert.

- **Subtropical** dry summer climate

  Summers are hot and dry, and winters are cooler and wetter.

- Humid subtropical climate

  Summers are hot and winters are cool; rain falls all year round.

- Humid oceanic climate

  Temperatures are cool, but not extremely cold, in winter and warm in summer, with rain all year round.

- Humid continental climate

  Summers are much hotter than winters; summers may have heavy rain and winters heavy snow.

- Subarctic climate

  Winters are long and bitterly cold. Most of the rain falls during summers, which are brief and cool.

- **Tundra** climate

  A tundra climate is dry; winters are very cold, and summers are not warm enough to melt the frozen water in the ground.

- **Icecap** climate

  Even in summer, temperatures may remain below the freezing point; there is no rain, though snow may fall.

- Highland climates

  Found in mountainous places, the highland climate changes depending on altitude (the height above Earth's surface); at the foot of a mountain, the climate may be tropical wet, while nearer the top it may be tundra.

A desert climate—rock formations in Monument Valley in the American Southwest.

EXTREME CLIMATES

Some climates are so extreme that they create **environments** in which it is almost impossible for anything to survive. For example, in Antarctica the climate is so cold and the land so icy that only a few plants, mostly mosses (small, green, nonflowering plants) can survive. Only a few tiny animals, such as insects, spend their whole lives on the mainland of Antarctica. The larger creatures that are found there, such as penguins, spend much of their lives off the coast, catching fish.

WHY CLIMATES VARY

Climates vary for many reasons, including **latitude** *(LAT uh tood)*, type of wind, distance from an ocean or a large lake, height above sea level, and what the surface of the land is like.

Latitude

The **equator** is an imaginary line around the middle of Earth. Distances north and south of it are measured in degrees of latitude. Places near the equator generally have warmer climates because the sun's rays shine directly onto them. Places farther away have cooler climates because the sun's rays are striking Earth at an angle and are less intense.

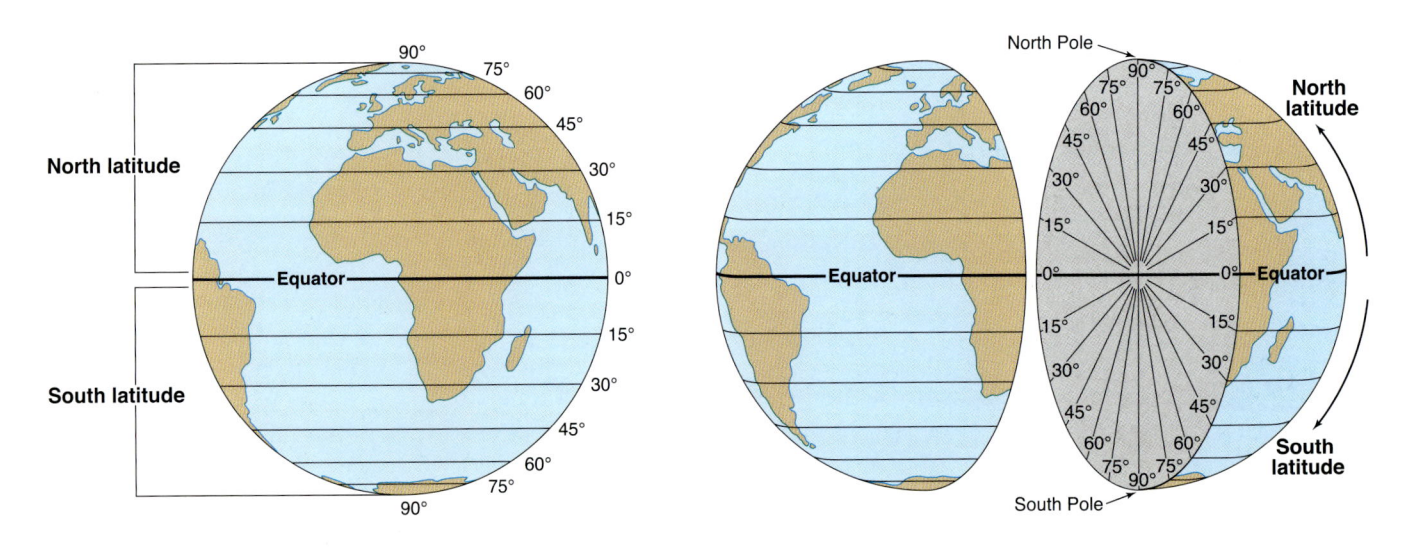

Points north and south of the equator are described in degrees of latitude. The equator is 0° latitude.

Wind

Wind is also an important factor in climate. Where the **prevailing wind** blows over water, such as a large lake or ocean, places near the shore will have a wet climate. This is because the wind is carrying water **vapor** that the air has absorbed from the ocean or lake.

Oceans and lakes

Water both warms up and cools down more slowly than land. Places near oceans and large lakes tend to have cooler summers than places inland. Since the water in summer is cooler than the land, the air is cooled over the water and this cool air then blows over the nearby land.

WHY CLIMATES VARY   17

In winter, the water is warmer than the land. The water warms the air above it and as this blows over nearby land, it makes the land warmer too. For example, along the northwest coast of the United States and Canada, the prevailing wind from the Pacific Ocean is warm and wet. Summers are cooler on the coast than inland, but winters are warmer.

The rain shadow

Mountain ranges affect **precipitation**. Winds blowing towards mountains are forced upward. This is known as the **windward** side of a mountain. On this side, the air is cooled and clouds develop, often resulting in rain or snow. As wind moves down the other side of the mountain (the **leeward** side), the air warms again and the clouds disappear. We call this dry area on the leeward side of mountains the rain shadow. Many of the world's **deserts** lie in rain shadows.

MICROCLIMATES

A **microclimate** occurs in a small, specific area. For example, a garden with walls to protect it from the cold wind may have a warmer microclimate than the surrounding area. Cities have microclimates. Because buildings and streets absorb large amounts of heat and then radiate that heat back again, cities are usually warmer than the countryside around them.

The rain shadow in the Pacific Northwest as seen in a photo taken from space. To the west (left) of the snow-covered Cascade Mountains (center), the lush vegetation is an indication of high levels of precipitation. The region to the east (right) of the mountains is in the rain shadow and receives less rain.

18 FORCES OF NATURE

CLIMATE AND THE ENVIRONMENT

The **climate** of a place helps to create its **environment**. All over the world, people, animals, and plants have adapted to the climate in which they live.

Animals

In Africa and India, many farmers keep zebu *(ZEE byoo)* cattle. These cattle are well adapted to hot **weather** because they perspire more than other kinds of cattle, helping to keep their bodies cool. In contrast, Highland cattle from northern Scotland have thick, shaggy coats that keep them warm, and so they are well adapted to cold weather.

Wild animals are usually well adapted to their climate. In places where winters are very cold, some animals spend many months in a kind of deep sleep called **hibernation** *(HY buhr NAY shuhn)*. During this time they use very little energy and so do not need to look for scarce food while temperatures are low.

A husky dog is well adapted for snow and cold. It has a thick outer coat and a downy undercoat that insulate it from the cold.

Plants

Climate affects the crops people grow. In **tropical** climates, farmers may grow pineapples and sugarcane, which will not survive freezing temperatures. In cooler places, farmers may grow apple or cherry trees, or winter wheat, which thrive in these conditions.

Some kinds of plants have adapted to particularly harsh climates. One example, the baobab *(BAY oh bab)* tree, is found in hot, dry areas of Asia and Africa. This tree can store up to 30,000 gallons (114,000 liters) of water in its trunk. The baobab uses this water to survive in times of low rainfall. In another example, pine trees grow in huge numbers in the cold regions of such countries as Canada, Russia, and Sweden. The needlelike leaves of these trees have a thick, waxy covering that protects them from low temperatures. In addition, pine **sap** contains a substance that prevents it from freezing.

HOMES FOR DIFFERENT CLIMATES

Over many centuries, people have found ways to build houses that are well suited to their climates. Houses made from sun-dried bricks, called adobe *(uh DOH bee)*, are found in the hot, dry countries of the Middle East, North Africa, and in parts of Central America and the southwestern United States. The thick walls of an adobe house absorb heat very slowly during the day, so the temperature inside remains comparatively cool. When night falls, the warm bricks help to keep the inside of the house at a comfortable temperature.

The trunk of the baobab tree can store large amounts of water, allowing the tree to live in arid climates.

OCEAN CURRENTS

Huge **currents** of water flow through the oceans. For example, the Gulf Stream is a current of warm water flowing through the Atlantic Ocean from the Caribbean to northern Europe. Because of the Gulf Stream, such countries as Norway and the United Kingdom have milder winters than other regions at the same **latitude**.

El Niño

Another warm ocean current, El Niño *(ehl NEEN yoh),* flows southward along the Pacific coast of South America, usually around Christmastime in late December. In fact, the El Niño current is named for the time at which it occurs. El Niño is Spanish for "the boy child," a phrase often used to refer to the Christ child in Spanish-speaking countries. Every two to seven years, this current is stronger than usual, affecting **climates** across the world. Scientists now use the term El Niño to describe all the changes in the ocean and the **weather** that are caused by this event.

An El Niño event

Usually the water in the Pacific Ocean is warmer in the west. The air above this water is also warmer and rises, creating an area of low pressure. Winds blow from South America toward the area of low pressure, carrying with them warm water from the ocean surface. Cold water, rich in **minerals** and the tiny life-forms on which fish feed, rises from deep in the ocean to replace the warm water, making the coastal waters of the Pacific Ocean off South America one of the most important fishing areas in the world.

Wildflowers bloom in a California desert after rainfall in an El Niño year. During an El Niño event, weather patterns can be affected worldwide.

OCEAN CURRENTS   21

When there is no El Niño event, the wind and ocean currents move from east to west.

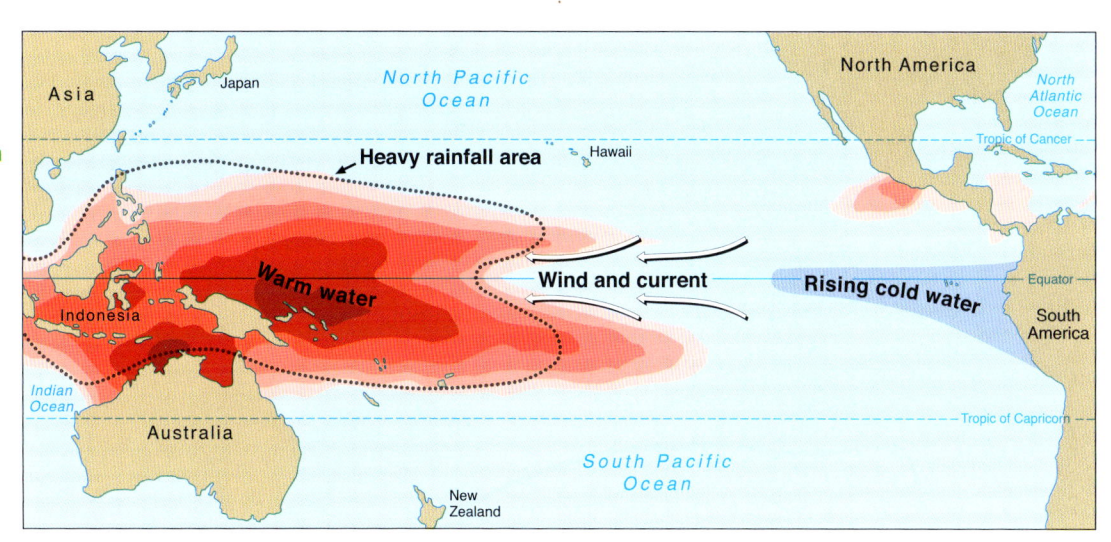

[Map showing normal conditions: Warm water and Heavy rainfall area over Indonesia/western Pacific, Wind and current flowing east to west, Rising cold water off South America]

During an El Niño event, the east-to-west winds and currents over the tropical Pacific weaken, or even reverse. The area of cold water off the coast of South America disappears almost completely.

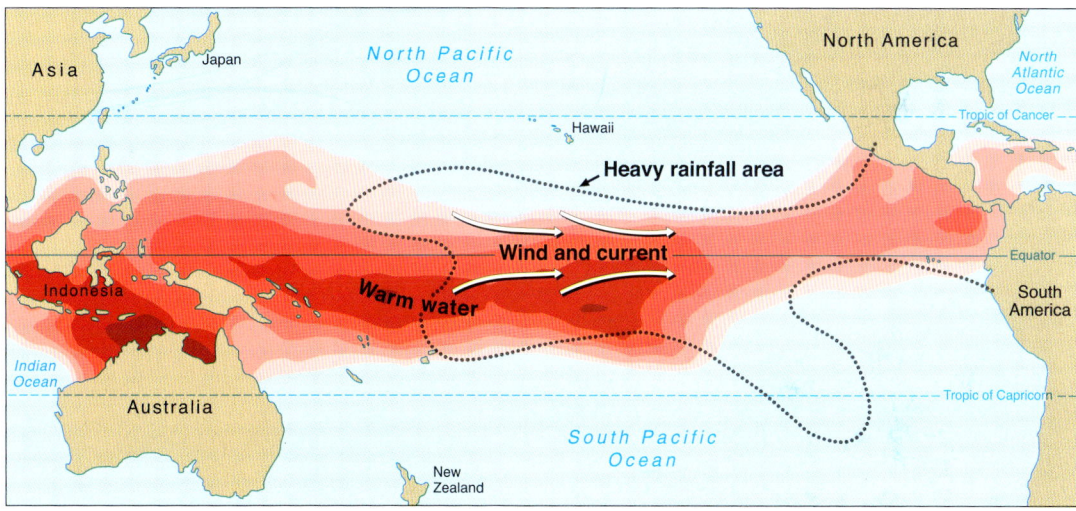

[Map showing El Niño conditions: Warm water extending across Pacific, Heavy rainfall area shifted eastward, Wind and current flowing west to east]

During an El Niño event, **atmospheric pressure** on the western side of the Pacific Ocean is higher than normal. Winds no longer blow warm water away from South America. They may even reverse, blowing warm water east. Cold water no longer rises to the surface along the South American coast; without their supply of food, the fish die or move away. The unusually low pressure brings heavy rain and often **floods** to the eastern Pacific, and Indonesia and Australia get less rain than usual.

LA NIÑA

A typical El Niño event lasts about 18 months. Afterward, weather patterns may return to normal, but often there is a period known as La Niña *(luh NIHN yuh)*, which means *"the girl."* La Niña is the opposite of El Niño. During La Niña, winds blowing across the Pacific Ocean from South America force even more warm water than usual into the western Pacific, leading to more rain than usual in the west and much drier conditions in the east.

SEASONS

We divide the year into four periods called seasons: spring, summer, autumn, and winter. Each season lasts approximately three months.

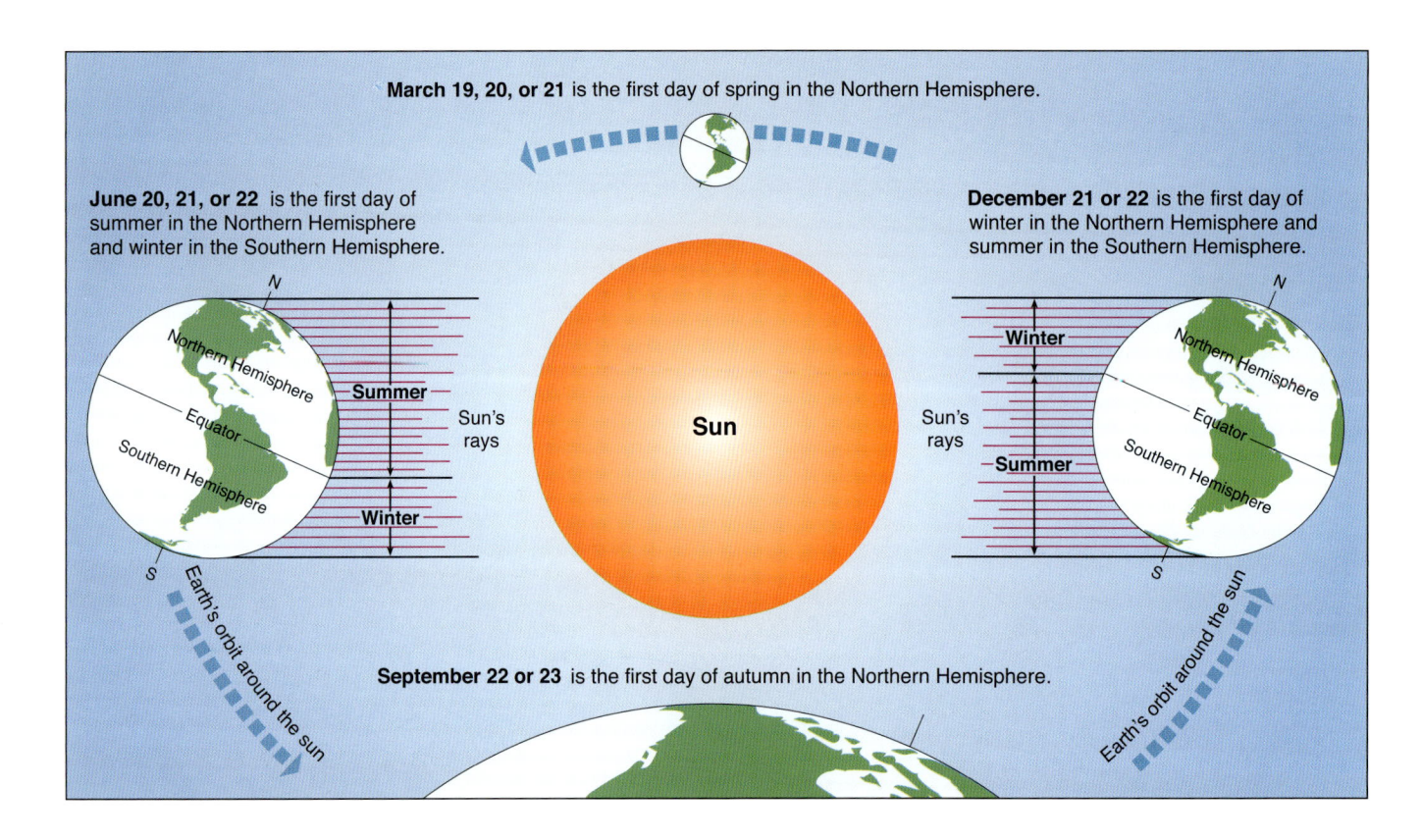

March 19, 20, or 21 is the first day of spring in the Northern Hemisphere.

June 20, 21, or 22 is the first day of summer in the Northern Hemisphere and winter in the Southern Hemisphere.

December 21 or 22 is the first day of winter in the Northern Hemisphere and summer in the Southern Hemisphere.

September 22 or 23 is the first day of autumn in the Northern Hemisphere.

Earth is tilted at around 23° on its axis. This causes different parts of Earth to receive different amounts of sunlight.

What causes the seasons?

We have seasons because of the changing position of Earth in relation to the sun. As it circles the sun, Earth is slightly tilted on its **axis**. Summer occurs in the Northern Hemisphere (the half of Earth that is north of the **equator**) when that part of Earth is tilted toward the sun. During this time, the sun's rays shine more directly onto the Northern Hemisphere. At the same time, it is winter in the Southern Hemisphere (the half of Earth that is south of the equator). Six months later, it is the Southern Hemisphere that is tilted toward the sun and it is summer in the Southern Hemisphere and winter in the Northern Hemisphere.

Planting and harvesting

In places with four seasons, spring is when farmers generally plant crops, which ripen over the summer. The crops are harvested at the end of summer or the beginning of autumn. Modern farming methods have changed the times during which some jobs are done, but farmers are still guided by the seasons.

Wet and dry seasons

Some places do not have four seasons. Around the equator, temperatures do not vary much through the year, so there is no "winter" or "summer." But, while there may not be much variation in temperature, in many places there are large variations in rainfall, giving wet and dry seasons. In the tropics, farmers plant their crops before the start of the wet season. The crops grow well as the rain falls; then crops ripen in the dry season. In southern Asia, the wet and dry seasons are caused by winds called **monsoons**. From June to September, the monsoon winds blow across the ocean from the southwest, bringing wet weather. From December to March, the monsoon winds reverse to blow from the northeast across the land, bringing dry weather.

FESTIVALS

All around the world, people celebrate the seasons with festivals. Hindus celebrate spring with Holi, a festival of color. During Holi, people throw colored water and powder at each other and light bonfires. Many festivals celebrate the season of autumn. In China, the Mid-Autumn Festival, or Moon Festival, is celebrated on the full moon that happens around the September equinox. At the Mid-Autumn Festival, families may gather to have a picnic that includes round "moon cakes," a baked pastry of thin crust filled with dense, sweet fillings.

A street in Varanasi, India, is flooded by monsoon rains, which begin in June and continue to September.

FORCES OF NATURE

UNSEASONABLE WEATHER

Sometimes temperatures are unusually high or low for the time of year. There may be no rain at a time that is usually wet, or there may be too much rain in months that are usually dry. These are examples of unseasonable **weather.** Weather that is too warm for the time of year can cause problems for both plants and animals. For example, buds may appear too early on trees and then be damaged by the cold when temperatures return to normal. Or, weather that is unseasonably cold may harm plants that are unused to below-freezing temperatures.

Volcanic eruptions

Unseasonable weather can happen for a number of reasons. One cause can be volcanic **eruptions.** When a **volcano** erupts, it often produces large amounts of ash, which are carried around the **stratosphere** by **jet streams.** The ash prevents some of the sun's heat from reaching the surface of Earth. For example, a volcano called Mount Tambora erupted in Indonesia in 1815. The following year, temperatures in the Northern Hemisphere were so cold that people called it "the year without a summer." Frosts and rains ruined crops, which led to severe food shortages. In another example, dust from the eruption of Krakatau (also in Indonesia) in 1883 is thought to have

The orange crop in California was damaged by unseasonably cold weather in 2007. Farmers choose crops to suit the climate, but they cannot always adjust for unseasonable weather.

lowered temperatures worldwide for the next five years. Ash and gases from the 1991 eruption of Mount Pinatubo in the Philippines also lowered temperatures globally.

Unreliable rains

In countries where farmers rely on seasonal rains, unseasonably dry weather can be disastrous. This can happen, for example, when the wet **monsoon** winds fail to arrive in southern Asia, often as a result of an El Niño event (see pages 20-21). Without rain, crops cannot grow, and there is nothing for people to eat the following year. It is just as bad if the rains fall for a while, then stop early, leaving newly planted crops to shrivel and die in the fields.

This sunset occurred after the 1991 eruption of Mount Pinatubo. Volcanoes emit particles into the stratosphere which can cause beautiful sunsets. These particles can also cause unseasonable weather.

UNSEASONABLE RAIN

If an area gets more rain than usual, the result may be **floods**. However, heavy rains during harvesttime, when farmers expect dry weather, can be damaging even if there are no serious floods. Such rains may make it impossible for farmers to harvest their crops, which then lie rotting in the fields.

CLIMATE CHANGE

The various regions on Earth have gone through many periods of **climate** change. Some places that have cool climates today were **tropical** forests millions of years ago. The Sahara, which today is desert, once had a wet climate that allowed grasses and trees to grow.

Ice ages and interglacial periods

There have been long periods in the history of Earth when the climate was much cooler than it is today. These periods are called **ice ages.** The most recent ice age ended about 11,500 years ago. During this ice age, sheets of ice covered nearly one-third of Earth. Huge moving masses of ice, called **glaciers,** wore away Earth's surface.

Between ice ages are periods of warmer weather, called **interglacial** *(IHN tuhr GLAY shuhl)* periods. We are in an interglacial period now, but even within this period, the climate has changed. About 1,000 years ago, Earth's climate went through a warmer period that lasted about 300 years. Then, from around 1400 to 1850, temperatures dropped in a period known as the Little Ice Age. During this time, glaciers in the Rocky Mountains, the Alps, and the Himalayas grew larger.

Angel Glacier, in the Canadian Rockies, is melting during the current warmer interglacial period.

CLIMATE CHANGE **27**

Layer of greenhouse gases

The greenhouse effect:
1) The sun's rays penetrate the atmosphere, warming Earth.

2) That heat is re-radiated from Earth's surface and is

3) absorbed by a layer of greenhouse gases (CO_2 and other gases).

4) Some of this absorbed heat is emitted into space.

5) A certain amount of this heat is re-radiated back to Earth's surface.

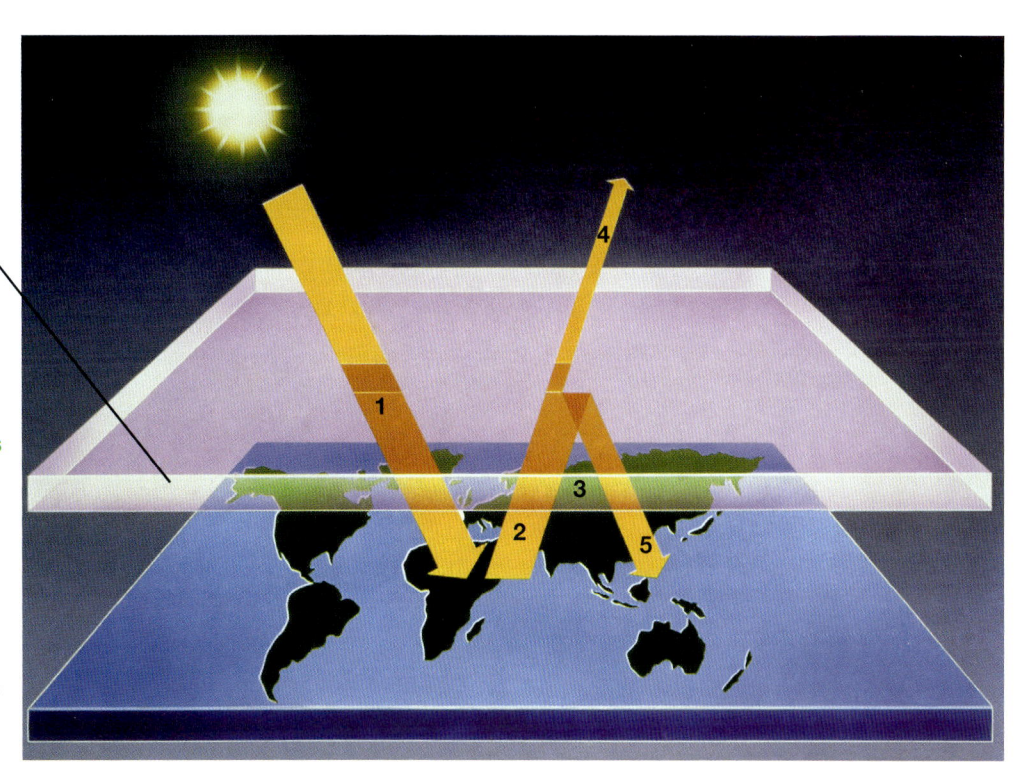

Global warming

Today, the climate seems to be warming up again. There are more spells of extremely hot weather, or **heat waves,** and more **droughts.** While deserts are becoming larger, glaciers are melting and shrinking. We call this change in Earth's climate **global warming.** Most scientists believe global warming is occurring because of the high levels of heat-trapping **greenhouse gases** being emitted into the **atmosphere** by human activities. These scientists fear that the increased temperatures could melt enough ice in the **Arctic** and **Antarctic** to cause sea levels to rise worldwide, leading to devastating flooding in many places.

THE GREENHOUSE EFFECT

Greenhouse gases (carbon dioxide, or CO_2, and other gases) in the atmosphere trap some of the heat energy from the sun in a process called the **greenhouse effect.** Without the greenhouse effect, Earth would be too cold for most living things to survive. Many scientists believe that today, due to human activity, there are too many greenhouse gases in the atmosphere, trapping too much heat. Scientists fear that by the year 2100, greenhouse gases could cause Earth's average temperature to rise by 2.5 to 10.4 °F (1.4 to 5.8 °C).

In the past, temperature changes on Earth have occurred over a much longer period of time. A rapid rise in average temperature in just 100 years would be a dramatic change that might lead to problems. For example, warmer temperatures would cause ice at the poles to melt and sea levels to rise, flooding coastal areas. Warmer ocean temperatures could also cause hurricanes and other storms to become more violent and more frequent.

FORCES OF GEOLOGY

EARTH'S LAYERS

At the center of Earth is a **core** of metal about 4,400 miles (7,100 kilometers) in diameter. Around the core is a layer of molten rock called the **mantle,** which is about 1,800 miles (2,900 kilometers) thick. Finally, there is an outer layer called the **crust,** on which we live. Nowhere on Earth is the crust more than 60 miles (100 kilometers) thick.

Plate tectonics

The land masses of the Earth, the **continents,** are constantly moving in relation to each other, sometimes by as much as 4 inches (10 centimeters) a year. Scientists have devised a theory, called **plate tectonics,** to account for this. They believe that the crust and the outer layer of the mantle are divided into about 30 rigid pieces, called **tectonic plates.** Some plates are huge: for example, most of the world's largest

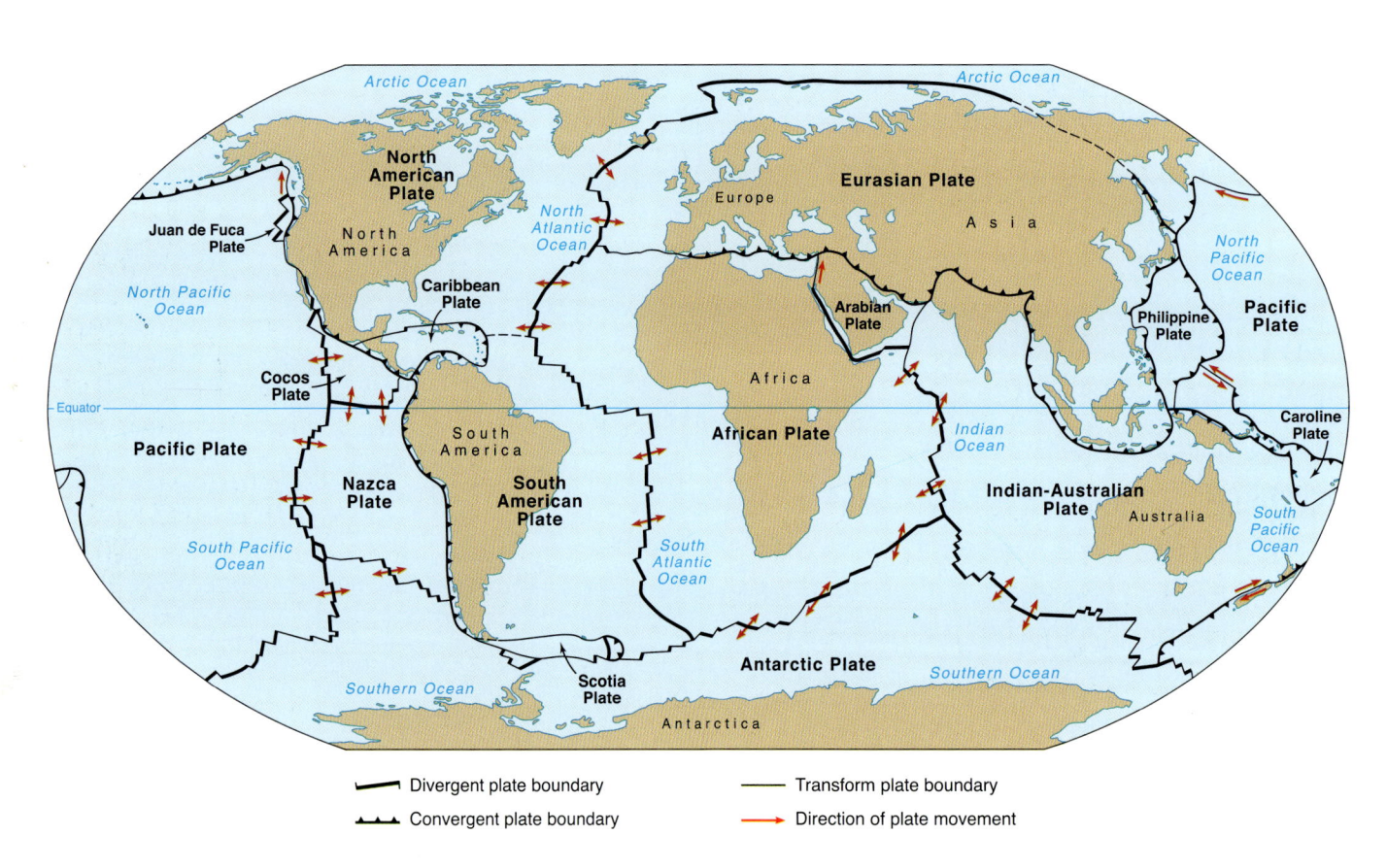

This map shows Earth's tectonic plates. Some plates, called convergent plates, are moving toward each other. Others, called divergent plates, are gradually pulling apart. In other places, called transform plate boundaries, plates are sliding alongside each other.

— Divergent plate boundary   — Transform plate boundary
⊥⊥⊥ Convergent plate boundary   → Direction of plate movement

ocean, the Pacific Ocean, sits on a single plate. Scientists believe that the tectonic plates move around on top of a layer of the mantle called the **asthenosphere** *(as THEHN uh sfihr)*.

In some places, the plates push toward each other, forcing one plate to sink below the other in a process called **subduction** *(suhb DUHK shuhn)*. This is what has happened at the plate boundary between India and the rest of Asia. This subduction has formed (and continues to form) the world's highest mountain system, the Himalaya.

In other places, plates slide alongside each other. In yet other places, such as under the Atlantic Ocean, the plates are drifting apart. Rock from the mantle beneath bubbles up into the gap, forming new crust.

As the plates move, so do the continents that sit on top of them. The result is that the face of Earth is slowly, but constantly, changing.

In California, the San Andreas fault (shown here in an aerial view) runs along part of the boundary between the Pacific and the North American tectonic plates.

THE MOVING CONTINENTS

Scientists believe that about 200 million years ago, almost all of the land on Earth was one big continent. This continent, called Pangaea *(pan JEE uh)*, gradually broke up to form the continents we know today. On a map of the world, it is still possible to see how the great landmasses could fit together like a giant jigsaw puzzle. A German scientist, Alfred Wegener *(VAY guh nuhr,* 1880-1930), noticed this and in 1912 theorized that the continents were moving. He called his theory continental drift. However, his ideas were not widely accepted by scientists until the late 1960's.

VOLCANOES

The word **volcano** comes from "Vulcan," the Roman god of fire and metalworking. The Romans believed that Vulcan's smithy was under Mount Etna, a volcano on the island of Sicily, which is off the coast of present-day Italy. A volcano is an opening in Earth's crust where ash, gas, and molten rock, called **magma,** is forced up to the surface, erupting through a hole, or vent. At the surface, ash and magma (called **lava** once it reaches the surface) often build up to create a mountain, which is also called a volcano.

A volcano builds up around a central vent. Most of the time, a volcano may appear to be inactive. But magma may be building up in the magma chamber beneath the surface. Eventually the magma may reach such great pressure that it bursts out of the mountain as rocks, ash, lava, and gases.

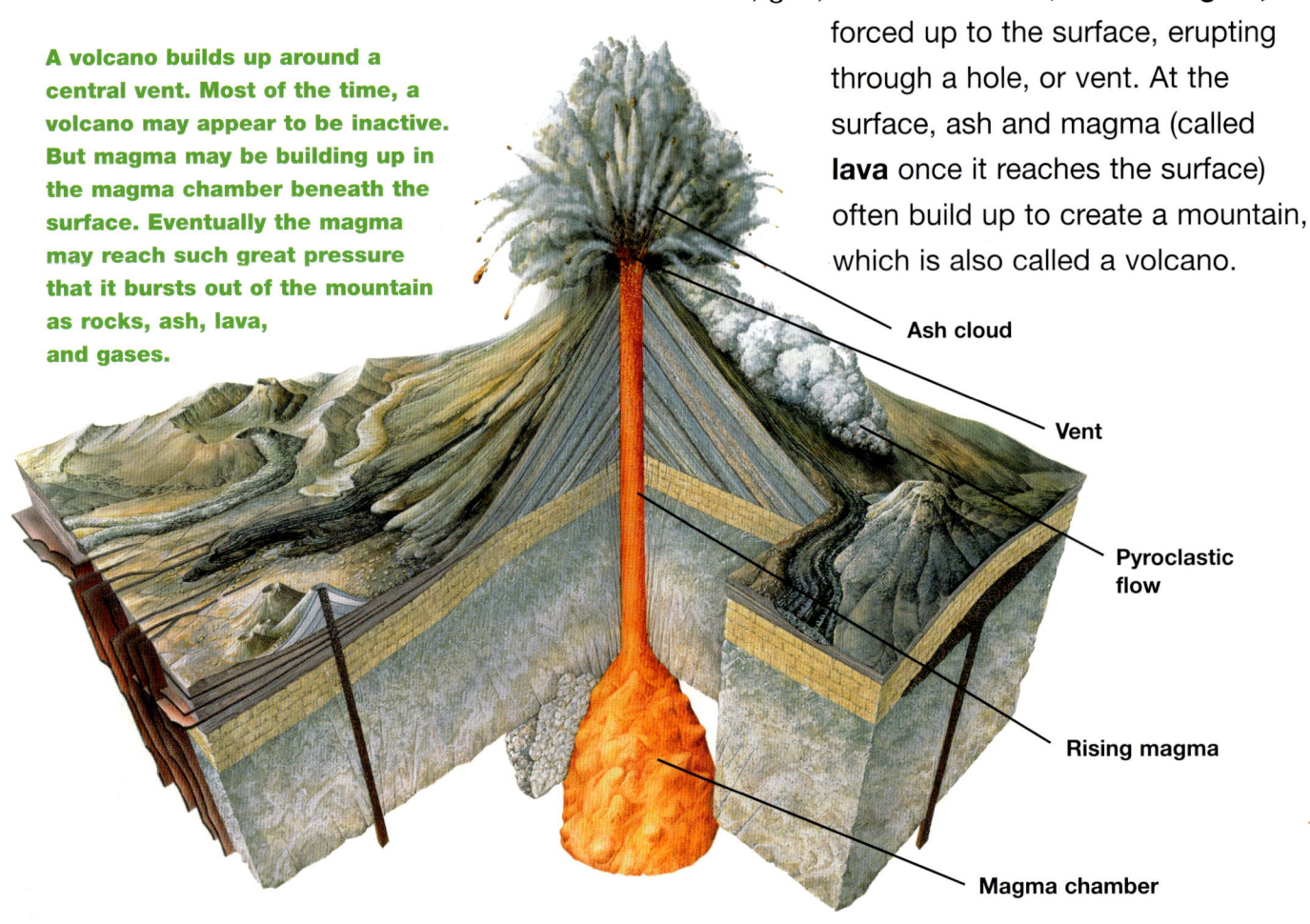

- Ash cloud
- Vent
- Pyroclastic flow
- Rising magma
- Magma chamber

The Ring of Fire

Most volcanoes occur along the edges of Earth's **tectonic plates**, usually where one plate is sinking down into the **mantle** beneath or at places where the plates are pulling apart. For example, at an area around the edge of the Pacific Ocean—known as the Ring of Fire—where the edges of Pacific and Nazca plates (see page 28) are pulling apart, there are a large number of volcanoes.

Some volcanoes occur above an underground area where there is a concentration of heat. These areas are known as hot spots. Some **hot spots** occur at the boundaries between plates, but some lie in the middle of plates.

Types of eruption

There are several different kinds of volcanic **eruptions.** Some eruptions produce thick, sticky lava that flows slowly down the side of the volcano; others put out faster-moving, thinner lava. Some eruptions primarily consist of gases, which can poison or suffocate people. Others hurl out showers of rocks, pebbles, or ash. The most dangerous eruptions occur when clouds of hot ash and gases pour at great speed down the side of the mountain, destroying everything in their path. These eruptions are called **pyroclastic** *(PY ruh KLAS tihk)* **flows.** Scientists believe that a pyroclastic flow from Mount Vesuvius in Italy in A.D. 79 destroyed the city of Pompeii and other nearby towns.

MOUNT PELÉE

In 1902, a volcano called Mount Pelée erupted on the Caribbean island of Martinique. The eruption took the form of a violent pyroclastic flow, which destroyed the city of St. Pierre. It is estimated that 28,000 people died as a result of this eruption. One of the survivors was a prisoner who was locked in an underground cell in the city's prison. By being below ground he was protected from the cloud of hot ash and gas.

The eruption of Mount Vesuvius completely buried the city of Pompeii in ash. The site lay almost untouched until the mid-1700's. Organized excavation began in 1860. Today, about three-fourths of the ancient city has been uncovered.

FORCES OF NATURE

EARTHQUAKES

Like **volcanoes,** most **earthquakes** occur along the edges of Earth's **tectonic plates.** They happen mainly at places where the plates are sliding past each other or colliding with each other. Most earthquakes are barely strong enough for people to feel, but some are extremely violent.

How does an earthquake happen?

The movement of the tectonic plates strains the rock at and near plate boundaries and produces zones of **faults** around these boundaries. Along the fault, large blocks of rock push against each other, building pressure. When the rock breaks, the blocks that had been unable to move are freed. The blocks suddenly slide against one another, and the energy released travels out in waves, called **seismic waves.** These seismic waves travel through the ground, causing it to shake. Most earthquakes start underground at a point called the **focus.** The point on the surface directly above the focus of an earthquake is known as the **epicenter,** and this is where the most violent shaking is usually felt.

The waves caused by an earthquake travel outward like ripples on a pond. As they travel, they become weaker, so places near the focus (see below) will experience more violent shaking than places farther away.

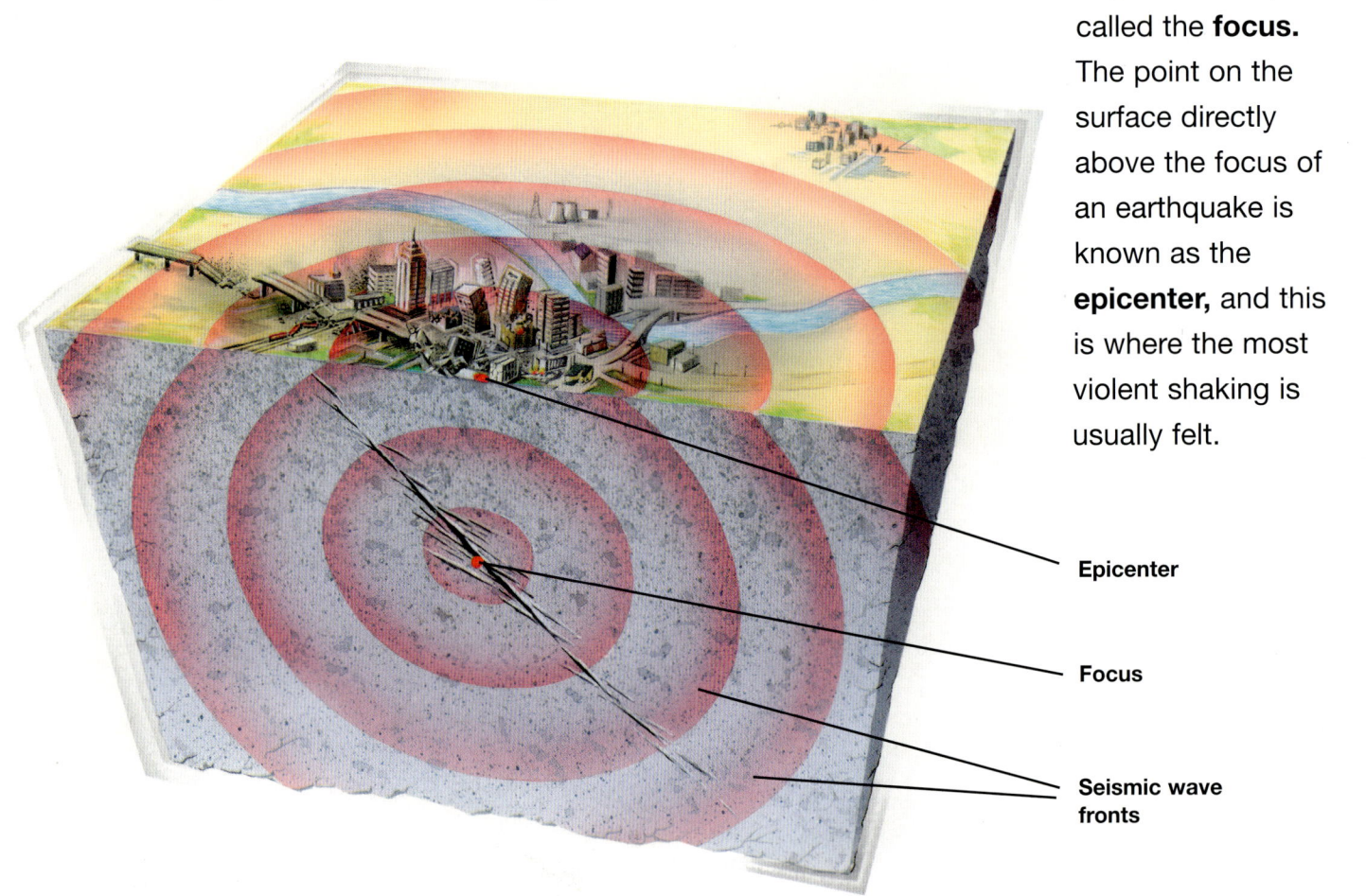

- Epicenter
- Focus
- Seismic wave fronts

Seismic waves

There are two kinds of seismic waves—body waves and surface waves. Body waves travel through Earth, causing rock to move in different directions. These are the fastest-moving and most damaging waves. Surface waves travel along Earth's surface. They move more slowly than body waves, and people usually feel little more than a slight rocking.

Earthquake hazards

During a serious earthquake, buildings and bridges may collapse, and trees and rocks may topple and fall. While earthquakes very rarely kill people directly, large numbers of people can be injured or killed by falling objects or collapsing buildings and structures. Another major hazard after an earthquake is fire, which can break out if power lines have fallen or gas pipes have broken. **Landslides** may block roads, making it difficult for rescue services to reach injured people or to bring food and clean water. This was one of the biggest problems after the earthquake that hit remote areas of India and Pakistan in October 2005, in which thousands of people died.

SCIENCE AND EARTHQUAKES

Scientists who study earthquakes are called **seismologists.** They measure earthquakes according to various scales. The first scale, the Richter, was developed in 1935 by American seismologist Charles F. Richter (1900-1985). Each point on the Richter scale marks an earthquake 32 times stronger than the last point—so an earthquake that measures 5.0 on the Richter scale releases 32 times as much energy as one that measures 4.0. To measure the largest earthquakes, seismologists now use another system, the moment magnitude scale (magnitude means *greatness of size*). Moment magnitude and Richter magnitude are about the same for earthquakes up to magnitude 7.

Survivors of the devastating 2005 earthquake in India and Pakistan cross a badly damaged bridge.

TSUNAMIS

A **tsunami** is a series of powerful ocean waves. Volcanic **eruptions** or undersea **earthquakes** often cause tsunamis. The word *tsunami* comes from two Japanese words meaning "harbor" and "wave."

Tsunami waves

In the middle of an ocean, tsunami waves are long and low. While most ordinary waves take only a few seconds to rise and fall, tsunami waves can take up to an hour. Because they are so low, however, sailors on board ships may not even notice these waves. In deep regions of the ocean, tsunami waves can travel at great speed, sometimes as fast as 600 miles (970 kilometers) per hour. When a tsunami reaches shallow areas near a coast, it slows down to about 20 to 30 miles (30 to 50 kilometers) per hour. As this happens, the water "piles up"—sometimes 100 feet (30 meters) or more above normal sea level. When these giant waves crash onto shore, they can flood as far inland as 3,500 feet (1,070 meters).

A disturbance deep under the ocean starts the tsunami.

Long, low waves travel at great speeds.

The waves "pile up" as they approach the shore.

Giant waves crash onto the shore.

Out at sea, tsunami waves are so low that ships may not even notice them. But once they reach shallower coastal areas these waves become much bigger and can be very destructive.

In some parts of the world, scientists have set up tsunami-detection systems that are designed to warn coastal communities when a tsunami is on its way. The systems measure the **seismic waves** of undersea earthquakes through sensors on the ocean floor. Other sensors measure changes in water pressure as the wave passes over the seabed.

The tsunami of 2004

In December 2004, an undersea earthquake near the Indonesian island of Sumatra triggered a giant tsunami that traveled across the Indian Ocean. It struck the coasts of other parts of Indonesia, as well as Sri Lanka, India, and Thailand. It even reached Somalia, about 3,000 miles (4,800 kilometers) from the tsunami's origin. Entire cities were destroyed in this tsunami and well over 200,000 people were killed.

ANIMAL SENSING

Some people believe that animals can sense when an earthquake or a tsunami is about to happen. People claim that just before the Indian Ocean tsunami of 2004, elephants in Sri Lanka began to scream and run to higher ground. Some scientists have suggested that animals may have a sharper sense of hearing than humans and that they can use this sense to detect when an earthquake is about to occur or a tsunami is coming.

Parts of Banda Aceh, on the Indonesian island of Sumatra, remain flooded fully one month after the city was devastated by a tsunami in December 2004.

LANDSLIDES, MUDSLIDES, AND AVALANCHES

In mountainous places, **landslides, mudslides,** and **avalanches** can all be hazards. In a landslide, loose rocks and soil slide down a mountain slope; mudslides are landslides of wet soil. An avalanche occurs when snow slides down a steep slope.

Landslides

Landslides can be caused by either heavy rain or **earthquakes**. Heavy rain can loosen rocks and soil so that they start to slip down a mountainside, gathering more rocks, stones, and soil as they fall. In 1983, almost the entire town of Thistle, Utah, disappeared under a lake that formed when a massive landslide blocked the Spanish Fork River. The water from the river flooded the town, destroying most of the buildings. In 1920, an earthquake triggered a huge landslide in

A man works to clear mud from his house in Guatemala after a mudslide in 2005.

LANDSLIDES, MUDSLIDES, AND AVALANCHES

Kansu Province in China. No one knows how many people died, but it could have been as many as 200,000.

Mudslides

Mudslides happen when wet soil becomes unstable. In February 2006, a series of mudslides in the Philippines buried an entire village in the southern region of the province of Leyte. Over 1,000 people died. The mudslide occurred after 10 days of heavy rain and a minor earthquake. Mudslides are particularly dangerous because there is little chance of anyone who is buried being able to find a pocket of air in the wet mud to allow them to survive until rescuers arrive.

Avalanches

An avalanche can take place when snow becomes unstable on a mountain. When a mass of snow breaks free, it slides down the mountain slope, burying everything in its path. Some avalanches can reach speeds of more than 100 miles (160 kilometers) per hour. An avalanche can be triggered by a small earthquake, strong winds, loud noises, or even by skiers.

When a mass of snow on a mountain breaks free, it can cause an avalanche, such as this one on Mount McKinley, in Alaska.

THE LOST TOWN OF PANABAJ

In 2005, heavy rains hit Guatemala in Central America. The mudslides that followed buried much of the town of Panabaj, in Guatemala's western highlands. About 1,000 people may have died. Some experts believe that the soil on the mountainsides surrounding the town became unstable because poor farmers had cleared the land of trees to grow crops. Trees help to prevent landslides and mudslides, because their roots hold soil in place and help absorb excess ground water.

FORCES OF DISEASE

MICROORGANISMS

There are many living things—called **microorganisms** *(MY kroh AWR guh nihz uhmz)*—that are too tiny to be seen except through a microscope. These life forms include **bacteria,** some kinds of **fungi,** and **viruses.** Some of these microorganisms are very useful to humans. For example, certain kinds of fungi and bacteria cause dead plants to rot, helping to enrich the soil. Bacteria are sometimes added to foods to change them. For example, bacteria are added to dairy products to create cheese and yogurt. Similarly, yeast is a kind of fungi that people use to make bread rise (puff up with air).

LOUIS PASTEUR

In 1865, Louis Pasteur began to research disease in silkworms. He noticed how infection could pass from one group of silkworms to another, and he became convinced that germs were to blame. He went on to develop **immunizations** *(IHM yu nuh ZAY shuhnz)* for animals and people against a whole range of diseases.

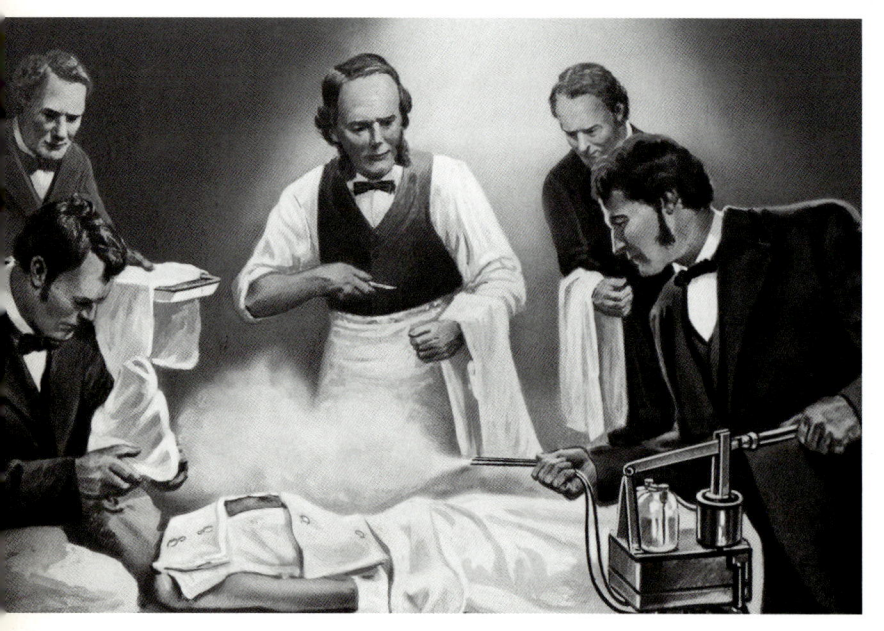

Germs

Some small life forms, however, can be incredibly dangerous. Microorganisms called **germs** cause diseases. Before the mid-1800's, when French scientist Louis Pasteur *(pas TUR,* 1822-1895) put forward his germ theory, there was little understanding of what caused diseases or how they spread. Hospitals were so dirty that many patients died from infections they caught during surgeries. Pasteur's work led a British surgeon, Joseph Lister (1827-1912), to realize how important it was to carry out surgery in sterilized conditions—that is, with instruments and bandages treated to

A drawing illustrates Joseph Lister directing his assistant in spraying antiseptic. Through studying the writings of Pasteur, Lister came to realize the importance of sterile conditions in surgery.

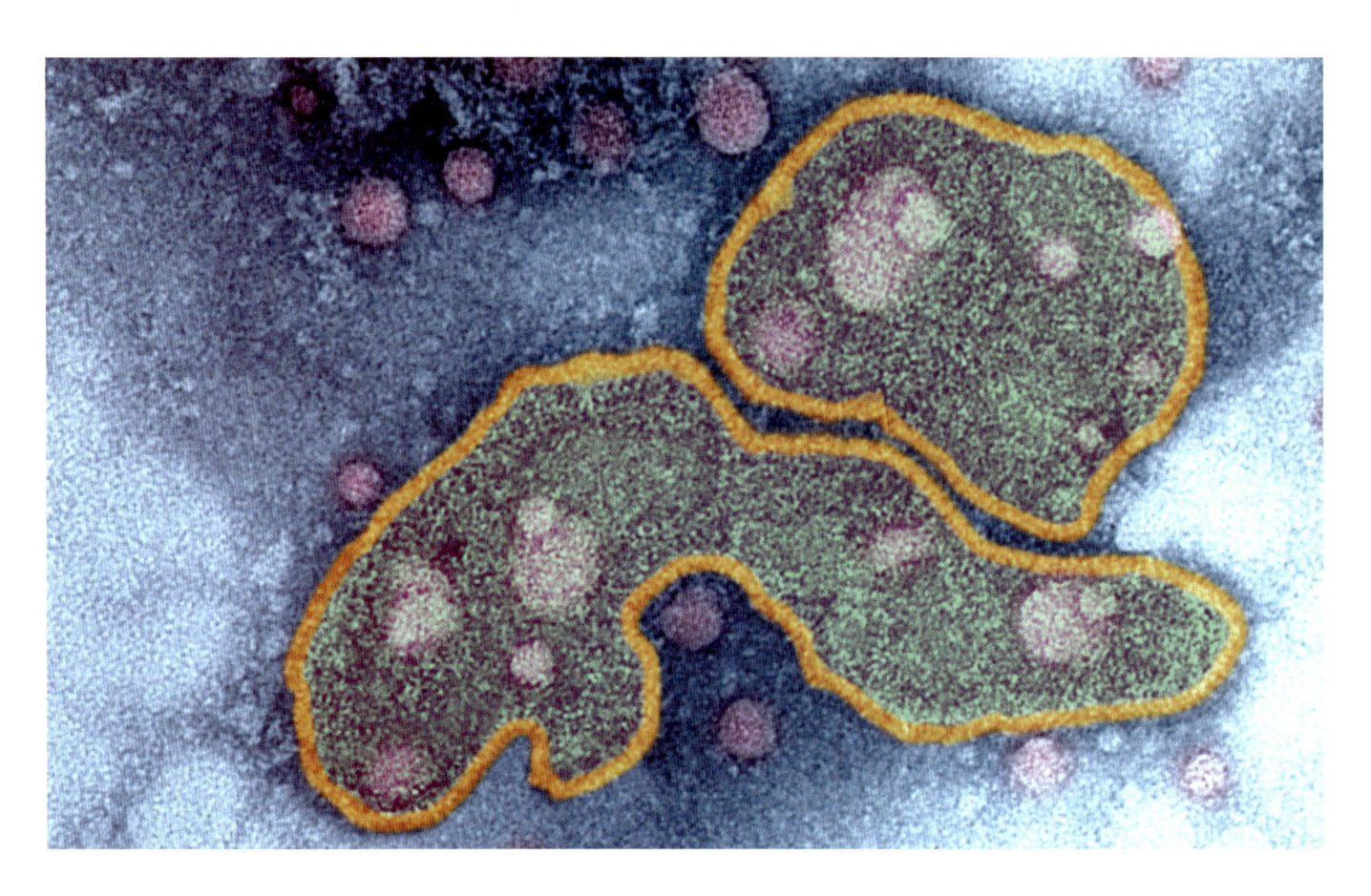

Two particles of a virus that causes influenza in birds are shown, highly magnified.

kill any germs. Today, understanding germs and how our bodies fight them are important parts of medicine.

Germs cause many diseases, from mild colds to serious illnesses, such as **AIDS** (acquired immunodeficiency syndrome). Some germs mainly affect animals, but they may occasionally pass from animals to humans. One strain of an **influenza** (IHN flu EHN zuh) that kills birds, for example, can also affect humans who come into close contact with infected birds. Other germs kill the animals and plants we need for food. In the 1840's, many poor farmworkers in Ireland lived mainly on potatoes. When a fungal disease called "potato blight" destroyed the potato crop, people did not have anything to eat, and many starved to death. Others were so weak from lack of food that their bodies were unable to fight off such diseases as **typhus** or **cholera.** The Great Irish Famine of 1845-1850 killed about 1 million people and forced another million to emigrate.

BACTERIA AND VIRUSES

Bacteria are living things consisting of a single **cell.** They are so tiny they can only be seen through a microscope. Bacteria are all around us, and large numbers of them live inside the human body. Some bacteria are useful. For example, certain kinds of bacteria in our **intestines** help us to digest our food.

Bacteria

Some bacteria cause disease. They enter a living thing and reproduce by dividing over and over again. As they do this, they produce substances that damage or destroy living tissue. Some bacterial diseases are very dangerous, for example, **tuberculosis** *(too BUR kyuh LOH sihs)*, or TB, which attacks the lungs. Another disease is **tetanus** *(TEHT uh nuhs),* which can infect a body through open cuts. Once inside, the tetanus bacteria produce a kind of **toxin** *(TOK suhn)* that affects the victim's muscles. The victim may have problems swallowing and even breathing, sometimes causing death. Medicines called **antibiotics** *(AN tee by OT ihkz)* can be used to kill some bacteria and so cure the diseases they cause.

A micrograph of the bacterium that causes cholera, Vibrio cholerae.

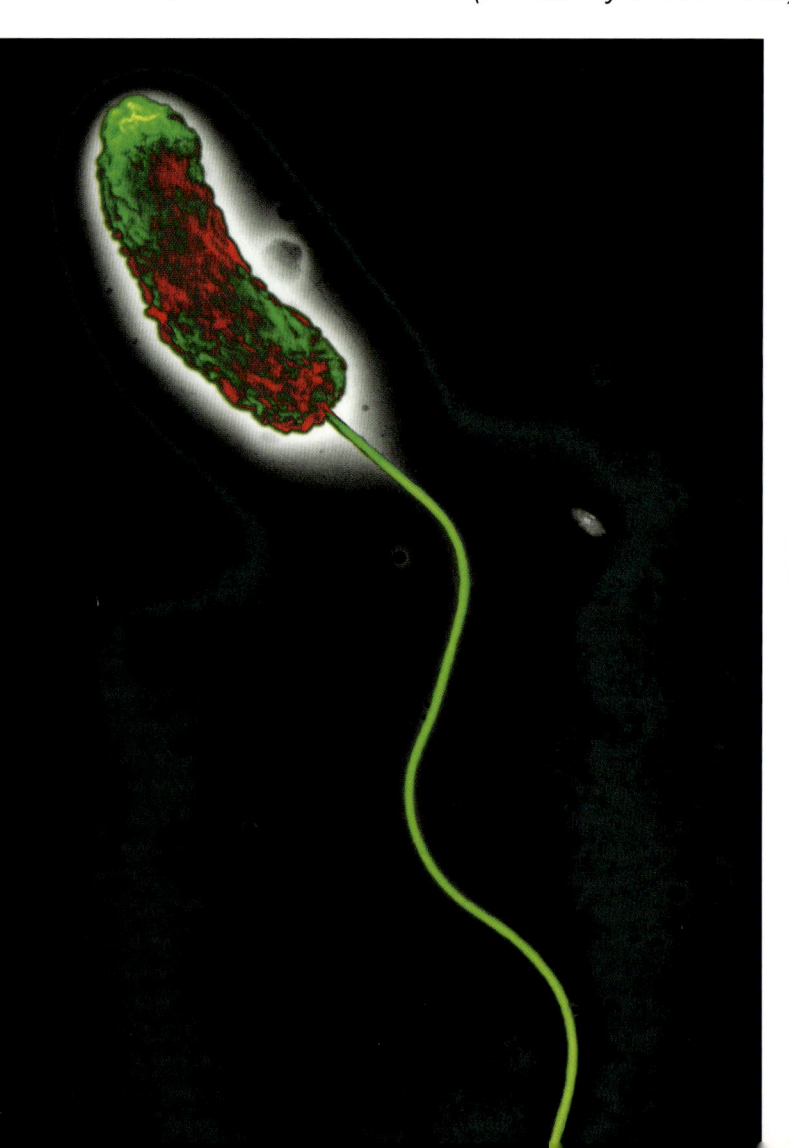

CHOLERA

One of the most dangerous diseases caused by bacteria—**cholera**—is common in places where people do not have access to clean water. The disease spreads when human waste from people infected with cholera gets into water and onto food. Cholera causes people to have severe **diarrhea,** and very often victims die from **dehydration.**

Viruses

Viruses are smaller than bacteria. A virus can reproduce only by entering the cell of another living thing. Once there, it multiplies and releases more viruses. Cells infected by viruses are either damaged or they die, resulting in disease. There are vaccines available to prevent diseases caused by certain types of viruses—for example, **poliomyelitis** (POH lee oh MY uh LY tihs) can be prevented with a vaccine. Antiviral medications can sometimes help to fight disease-causing viruses for which no vaccine exists. Often, however, for many viral diseases, all physicians can do is to treat the symptoms (signs) of a disease and wait for the body to fight the disease on its own. For example, the common cold is caused by a virus. There is no cure, but most people recover quickly.

Immunization

Many diseases can be prevented by **immunization**. This is usually done by injecting a weakened or harmless form of a bacterium or virus, called a vaccine, into the body. The vaccine does not infect the person with the disease, but it does trigger the body into making **antibodies** against the disease. If the person then comes into contact with a harmful form of the bacterium or virus, the antibodies protect the body against the disease.

American scientist Jonas Salk giving a child a polio vaccine. Salk developed the vaccine, which was the first effective weapon against polio, in the early 1950's.

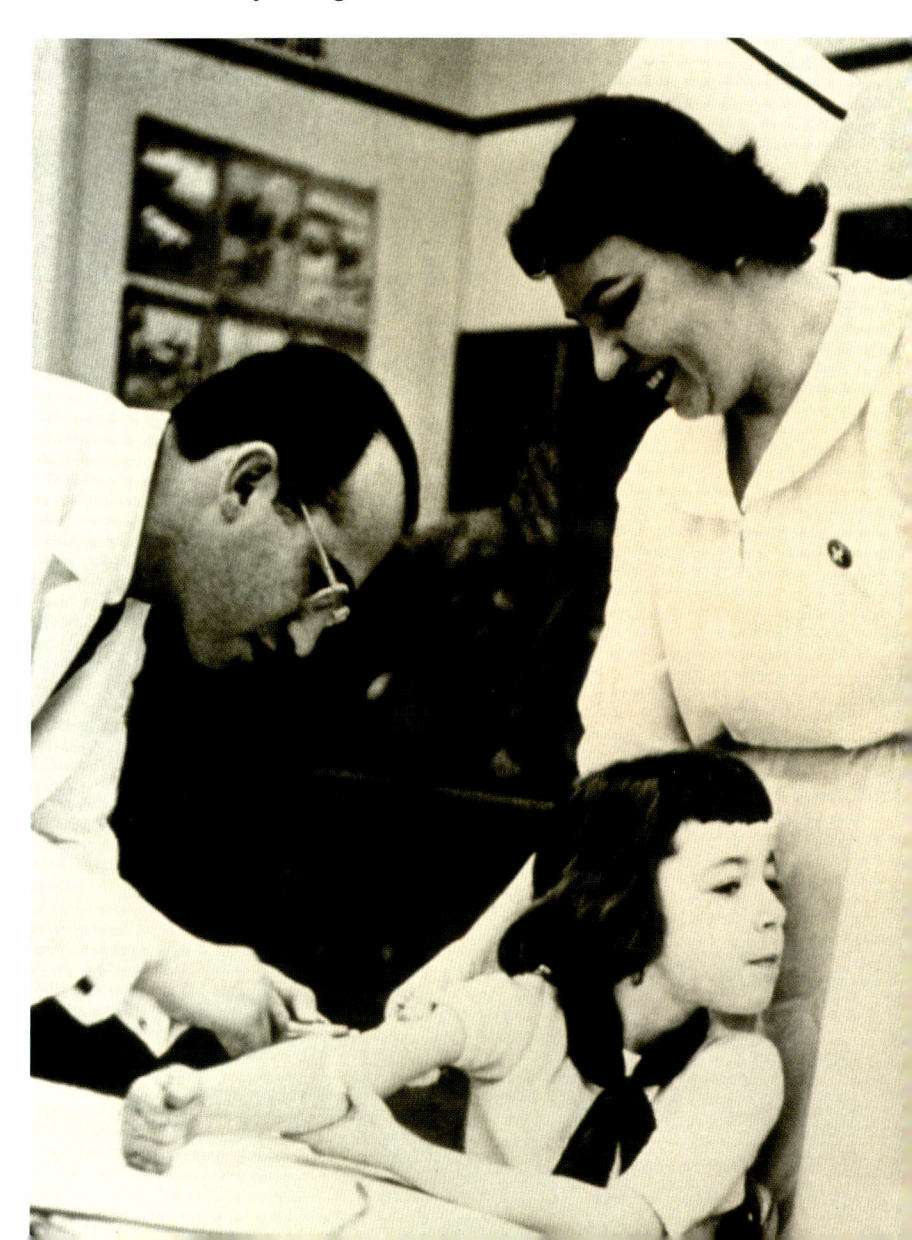

PARASITES AND FUNGI

Parasites are **organisms** (AWR guh nihz uhmz) that live and feed on or in other creatures. The creature on which a parasite lives is called its **host**. Some parasites cause little harm, but others can damage or even kill their host.

Dangerous parasites

Fleas, lice, and ticks are all examples of parasites. Such parasites bite their hosts to feed off the hosts' blood. These bites make the host's skin itchy and sore. Parasites can also spread dangerous **germs,** such as those that cause the disease **typhus.**

Some kinds of worms live as parasites in the digestive system of people or animals. These parasites absorb the food the host has consumed as it passes through the digestive tract, with the result that the host does not get the nourishment it needs. Another type of worm, the hookworm, is dangerous because it feeds on blood in a host's **intestines,** making the host very weak.

Worms or their eggs are often spread from person to person or animal to animal through food or water that is contaminated with **feces** (FEE seez). In **tropical** regions, worms called schistosomes (SHIHS tuh sohmz) are often found in fresh water. These worms spend part of their lives as parasites in snails before they enter the water. If a person swims or wades in water infected

Freshwater snails, such as the type shown here, can often be infected with parasitic worms called schistosomes.

PARASITES AND FUNGI

with schistosomes, the worms may burrow into his or her skin. The worms enter the person's bloodstream and can cause a serious disease called schistosomiasis (SHIHS tuh soh MY uh sihs).

Malaria

Malaria kills more than 1 million people every year. This disease is caused by tiny parasites called plasmodia *(plahz MOH dee uh),* and it is spread when a female *Anopheles (uh NOF uh leez)* mosquito bites someone whose body contains these parasites. The plasmodia enter the mosquito's body and reproduce. When the mosquito bites someone else, it passes on the plasmodia to a new host. The plasmodia enter the liver and blood of the new host, multiplying and destroying cells and causing high fever, headache, muscular pain, and nausea *(NAW shuh).*

FUNGI

Few types of fungi are deadly to humans. In fact, we eat fungi when we eat mushrooms. But, certain fungi can lead to terrible diseases. Ergot *(UR guht)* is a parasitic fungus that attacks grasses. Ergot attacks the grain of such plants as wheat, barley, and rye. If flour is ground from plants infected with ergot, people can get a disease called ergotism from bread baked from this flour.

Ergotism reduces the circulation of the blood, so gangrene *(GANG green),* the death of body tissue from lack of oxygen, can result in people infected with the disease. Gangrene can lead to amputation of limbs. Ergotism can also cause insanity, convulsions, and death. In earlier times, entire villages could be infected with ergotism when flour from the local mill was made from infected grain.

The best protection against malaria is to avoid mosquito bites. A mosquito net helps to prevent people being bitten while they are asleep.

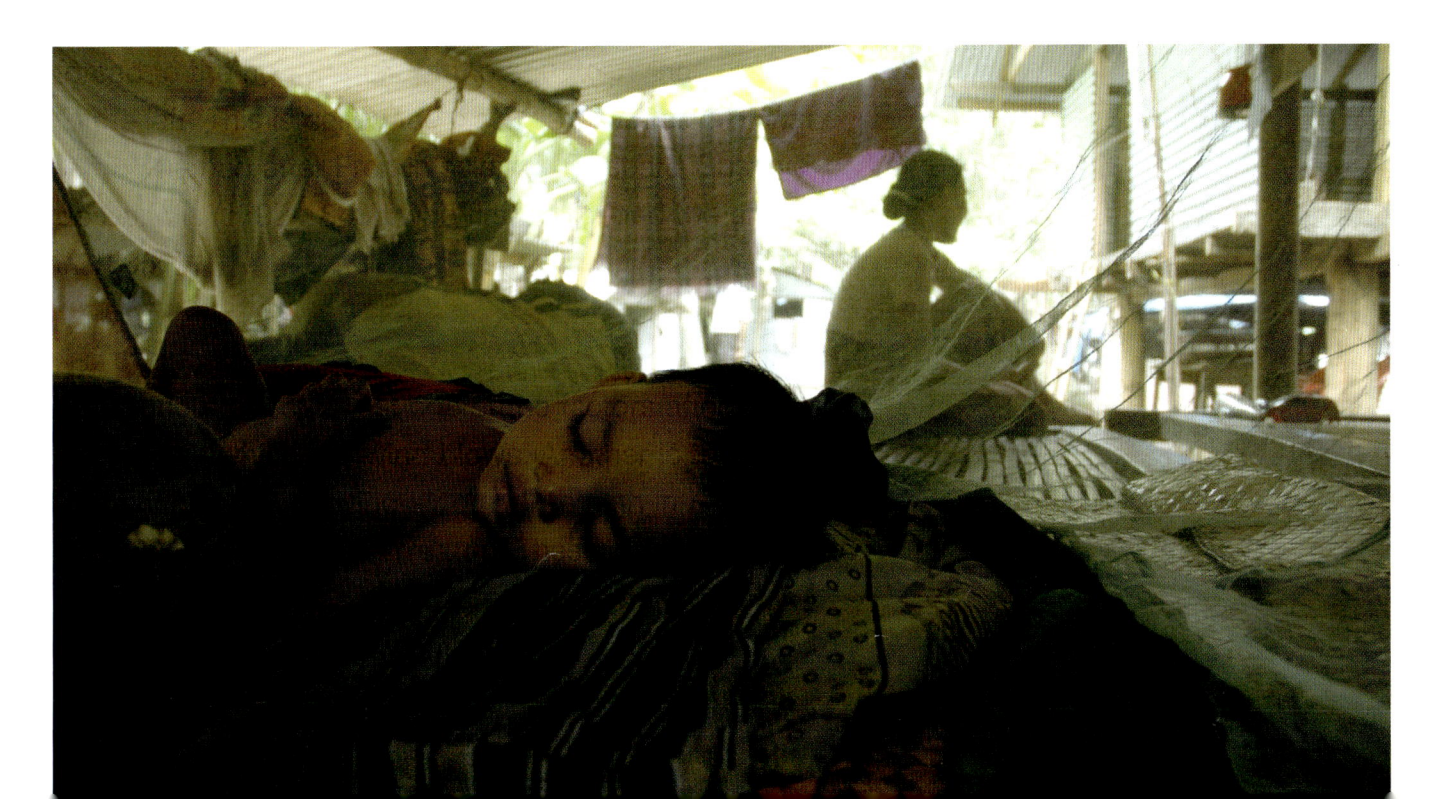

ACTIVITY

RECREATE PANGAEA

About 200 million years ago, almost all of the land on Earth was one big **continent**. There were no people around then to give it a name, but scientists today refer to this landmass as Pangaea *(pan JEE uh)*. Slowly, Pangaea broke up, and the pieces drifted apart. This activity helps you see how the continents that make up today's Earth were once part of Pangaea.

Equipment
- Map of the world (on this page)
- Tracing paper
- Pencil

Instructions

1. Using a photocopy machine that can enlarge images, photocopy the bottom half of this page at 150 percent.

2. Glue this copied image onto a piece of thin cardboard.

3. Now cut along the outlines of the pieces and push them together to make a shape like the continent of Pangaea. You can use the map of Pangaea (left) as a guide.

Pangaea (above) broke up to form the continents that make up today's Earth (below).

GLOSSARY AND ADDITIONAL RESOURCES

AIDS Also called acquired immunodeficiency syndrome, is the final, life-threatening state of infection with the human immunodeficiency virus (HIV).

Antarctic The region at and around the South Pole.

antibiotic A substance produced by living things, such as fungi, that can be used to destroy harmful bacteria.

antibody A substance made by human beings and animals to fight off foreign substances that invade the body, such as viruses and bacteria.

Arctic The region at and around the North Pole.

asthenosphere A part of Earth's mantle formed of a layer of hot rock.

atmosphere The layer of gases surrounding Earth.

atmospheric pressure The weight of the air pressing down on Earth's surface.

avalanche A mass of snow and ice that slides down a mountain slope.

axis An imaginary straight line running through the center of Earth from the North to the South pole.

bacterium (plural: **bacteria**) A living thing made up of a single cell. Some bacteria cause diseases, while others are helpful to human beings.

blizzard A heavy snowstorm with high winds.

cell The basic unit of all living matter. All living things are made up of cells.

cholera A water-borne disease caused by bacteria.

climate The average weather in an area over a period of time.

condense To change from a gas to a liquid as a result of cooling.

continent A large landmass. There are seven continents on Earth: Africa, Antarctica, Asia, Australia, Europe, North America, and South America.

core The center part of Earth's interior, lying below the mantle.

Coriolis effect The name for the effect that Earth's rotation has on the motion of anything moving over Earth's surface, including prevailing winds. South of the equator, prevailing winds are forced to the left by the Coriolis effect, north of the equator, to the right.

crust The solid outer layer of Earth.

current The movement of water or air in a particular direction.

dehydration A condition caused by a loss of fluids from the body.

desert A hot, barren region that receives little rainfall.

diarrhea Frequent and loose bowel movements.

drought A long period of unusually dry weather.

earthquake A shaking of the ground caused by the sudden movement of underground rock.

environment Everything that is outside a living being forms that being's environment.

epicenter The point on Earth's surface directly above the center (focus) of an earthquake.

equator An imaginary line around the middle of Earth, halfway between the North and South poles.

eruption The pouring out of gases, lava, and rocks from a volcano.

evaporate To change from a liquid into vapor.

fault In geology, a break in Earth's crust.

feces Solid waste matter from a human or animal body.

flood An overflow of water on land that is usually dry.

focus The point under Earth's surface where an earthquake starts.

front In weather, the place where two masses of air meet.

fungus (plural: **fungi**) Organisms that obtain food by absorbing nutrients from other living organisms or from parts of formerly living things.

germ A microorganism that causes a disease.

glacier A large mass of ice that flows, or moves, slowly by gravity.

global warming The gradual warming of Earth's atmosphere over many years.

greenhouse effect The warming of the lower atmosphere and surface of a planet by a complex process involving sunlight, gases, and particles in the atmosphere. On Earth, the greenhouse effect began long before human beings existed. However, recent human activity may have added to the effect.

greenhouse gas A gas that warms the atmosphere by trapping the heat from the sun that is reflected from Earth's surface.

hibernation An inactive, sleeplike state in an animal, during which its body temperature drops and its breathing becomes very slow.

host The living thing on which a parasite lives and from which the parasite gets its food.

hot spot An underground concentration of heat that creates volcanoes.

hurricane A tropical storm over the North Atlantic Ocean, the Caribbean Sea, the Gulf of Mexico, or the Northeast Pacific Ocean.

ice age A period in Earth's history when ice sheets cover vast regions of land.

ice storm A storm during which freezing rain falls onto cold surfaces and freezes instantly.

icecap A mass of ice and snow covering a large area.

immunization A way of protecting human beings and animals from disease using vaccines or serums to trigger the body to create antibodies to certain bacteria or viruses.

influenza An infectious disease caused by a virus; symptoms of influenza include fever and headaches.

interglacial A period between ice ages.

intestines The muscular tube in the body through which food and the products of digestion pass.

jet stream A fast-moving current of air high in the atmosphere.

landslide A mass of soil and rock that slides down a slope.

latitude The distance north or south of the equator, measured in degrees. The equator is 0°, the North Pole is 90° north, and the South Pole is 90° south.

lava Molten rock that flows out of a volcano.

leeward Facing away from the prevailing wind.

magma Molten rock beneath Earth's surface.

malaria A disease caused by tiny parasites called plasmodia, which are spread by the female *Anopheles* mosquito. Malaria causes a high fever, headache, muscular pain, and nausea, and can be fatal.

mantle The layer of rock between Earth's crust and core.

mesosphere The layer of Earth's atmosphere between the stratosphere and the outer layer, the thermosphere.

microclimate The climate of a small, specific area, which is different from the prevailing climate around it.

microorganism A living thing too small to be seen except with a microscope.

mineral A nonliving substance found in nature, the atoms of which are arranged in a regular pattern, forming solid units called crystals.

monsoon A seasonal wind that blows across the Indian Ocean and surrounding land areas.

mudslide Landslides of wet soil, or mud.

organism An individual plant or animal.

parasite An organism that feeds on and lives on or in another living thing, called the host.

planet A round body that orbits a star.

plate tectonics A theory that explains the origin of most of the major features of Earth's surface. The explanation involves the movement of rigid pieces, or plates, that make up Earth's outer layer.

poliomyelitis A serious viral infection in human beings that can lead to paralysis.

precipitation Moisture that falls from clouds, such as rain, snow, or hail.

prevailing wind The usual wind experienced in a particular place.

pyroclastic flow A cloud of hot ash and gas that travels at great speed mostly along the ground.

sap The liquid found in the stems, roots, and leaves of plants.

seismic waves Vibrations that travel through Earth, often caused by movement along a fault or volcanic eruptions.

seismologist A scientist who studies earthquakes.

semiarid Having very little rainfall.

stratosphere The layer of Earth's atmosphere directly above the lowest layer, the troposphere.

subduction When the edge of one of the tectonic plates that makes up Earth's surface sinks below a neighboring plate.

subtropical Bordering the tropics.

tectonic plate One of about 30 rigid pieces making up Earth's surface.

tetanus A disease that affects muscles, which is caused by toxins (poisons) produced by bacteria.

thermosphere The layer of the atmosphere farthest away from the surface of Earth; this layer extends into space.

tornado A rapidly rotating column of air that forms under a thundercloud or a developing thundercloud.

toxin A poison produced by a living organism.

tropical To do with the tropics—regions of Earth that lie within about 1,600 miles (2,570 kilometers) north and south of the equator.

tropical cyclone A large, powerful storm that forms over tropical waters.

troposphere The layer of atmosphere closest to the surface of Earth.

tsunami A series of powerful ocean waves produced by an earthquake, landslide, volcanic eruption, or asteroid impact.

tundra A huge region of the Arctic where no trees grow and where the land beneath the surface remains frozen all year.

tuberculosis An infectious disease caused by a type of bacteria that mainly affects the lungs but can also involve other organs.

GLOSSARY AND ADDITIONAL RESOURCES

typhoon A tropical storm in the Northwest Pacific Ocean.

typhus Any one of a group of important diseases caused by rickettsias, a special type of bacteria.

virus A tiny organism that can reproduce or grow only by entering the cell of another living thing. Viruses cause many serious diseases.

volcano An opening in the crust through which ash, gases, and molten rock (lava) from deep underground erupt onto Earth's surface.

water cycle (also called the **hydrologic cycle**) The continuous movement of water as it evaporates from Earth's surface, rises into the air, cools, condenses back into water, and returns to Earth's surface.

water vapor A gas formed by heating water.

waterspout A tornado over an area of water.

weather The state of the atmosphere in a particular place at a particular time.

wildfire An uncontrolled fire on open land (especially a forest fire) that endangers lives and property.

windward Facing toward the prevailing wind.

ADDITIONAL RESOURCES

BOOKS

General

Earth's Spheres, by Gregory L. Vogt, 6 volumes, Lerner, 2006-2007.

Eyewitness Earth, by Susanna Van Rose, Dorling Kindersley, revised ed., 2005.

Guide to Savage Earth, by Trevor Day, Dorling Kindersley, 2001.

Nature's Fury: Eyewitness Reports of Natural Disasters, by Carole Garbuny Vogel, Scholastic Reference, 2000.

Forces of Weather

Do Tornadoes Really Twist? Questions and Answers About Tornadoes and Hurricanes, by Melvin and Gilda Berger and Higgins Bond, Scholastic, 2000.

The Earth's Weather, by Rebecca Harman, Heinemann Library, 2005.

Eyewitness: Hurricane and Tornado, by Jack Challoner, Dorling Kindersley, revised ed., 2004.

Forces of Geology

Navigators: Volcano, by Anne Rooney, Dorling Kindersley, 2006.

Tsunami: The World's Most Terrifying Natural Disaster, by Geoff Tibballs, Carlton, 2005.

Volcanoes and Earthquakes, by Ken Rubin, Simon & Schuster, 2007.

Forces of Disease

Fighting Infectious Diseases, by Robert Snedden, Heinemann Library, revised ed., 2007.

Pox, Pus, and Plague: A History of Disease and Infection, by John Townsend, Raintree, 2006.

When Birds Get Flu and Cows Go Mad! How Safe Are We? by John DiConsiglio, Franklin Watts, 2007.

WEB SITES

http://bt.cdc.gov/disasters
http://earthobservatory.nasa.gov/
http://www.aoml.noaa.gov
http://www.bbc.co.uk/science/hottopics/naturaldisasters/
http://www.fema.gov/kids/index.htm
http://www.srh.noaa.gov/fwd/glossarymain.html

INDEX

Page numbers in *italics* indicate pictures

animals 7, 13, 15, 18, 35, 39
antibiotics 40
asthenosphere 29
atmosphere 6, 7, 10, 27
atmospheric pressure 10, 21
avalanches 36, 37

bacteria 5, 38, 40, 41
baobab trees 19

cholera 5, 39, 40
climate 9, 14-15, 16-17, 18-19, 20–21, 26–27
  and environment 18-19
  change 26–27
  map of world climates 14
clouds 6, 7, 8, 9
continents 28, 29
Coriolis effect 11
crops 7, 19, 23, 24, 25, 37, 39, 43
crust (Earth's) 5, 28, 29

deserts 9, 12, 14, 15, 17, *20*, 26
diseases 5, 38, 39, 40, 41, 43
droughts 7, 27

Earth 4, 5, 22, 28–29
  axis 10, 22
  climates 14–15
  temperature 9, 10, 16, 26–27
earthquakes 32–33, 34, 35, 37
El Niño 20–21, 25
  maps of El Niño event 21
eruptions 5, 24, 25, *30,* 31, 34

farming 19, 23, 37, 39
faults *29,* 32
fish 13, 20, 21
floods 4, 7, 21, *23,* 25, 27, *35,* 36
fog 7, 9

frost 9, 24
fungi 38, 39, 43

general circulation 10, 11
germs 38–39
glaciers 26, 27
global warming 27
greenhouse effect 27
greenhouse gases 27

hail 6, 9
heat wave 27
hot spots 31
hurricanes *4, 10,* 12, 13, 27
  Katrina *4,* 13

ice ages 26
ice storm 4
immunization 41
influenza 39
interglacial periods 26

jet streams 11, 24

La Niña 21
landslides 33, 36–37
latitude 16, 20
lava 30, 31

magma 30
malaria 43
mantle 28, 29, 30
mesosphere *6,* 7
microclimate 17
microorganisms 38
mist 7, 9
moment magnitude scale 33
monsoon 23, 25
mudslides 36, 37

oceans 8, 9, 12–13, 16–17, 20–21, 34–35
  currents 9, 20–21

Pangaea 29, 44
parasites 42–43
plants 7, 15, 19, 24, 25, 38, 39, 43
plate tectonics 28
precipitation 6, 8, 9, 17
pyroclastic flows 31

rain 4, 6, 7, 8, 9, 14, 15, 17, 19, 21, 23, 24, 36, 37
rain shadow 17
Richter scale 33
Ring of Fire 30

seasons 22–23
seismic waves 32, 33, 35
snow 6, 7, 9, 12, 17, *18,* 36, 37
storms 4, 12–13, 27
stratosphere *6,* 7, 24
subduction 29

tectonic plates 28–29, 30, 32
  map of tectonic plates 28
thermosphere *6,* 7
thunderstorms 9
tornadoes 4, *5,* 13
tropical cyclones 12, 13
troposphere *6,* 7
tsunamis 34–35
typhoons 12

vaccines 41
viruses 38, 39, 41
volcanoes 5, 24, 25, 30–31, 32

water cycle 8–9
waterspouts 13
weather 4, 6–7, 11, 14, 18, 24
wildfire 4
winds 6, 10–11, 12-13, 16, 17, 20, 21, 37
  map of prevailing winds 11
  prevailing 11, 16, 17
  trade 11